THE DATA MANAGEMENT WORKBOOK

The Data Management Workbook

Practical Exercises for Better Organization,
Storage and Use of Your Research Data

Kristin Briney

Pelagic Publishing | www.pelagicpublishing.com

First published in 2026 by
Pelagic Publishing
20–22 Wenlock Road
London N1 7GU, UK

www.pelagicpublishing.com

The Data Management Workbook: Practical Exercises for Better Organization, Storage and Use of Your Research Data

https://doi.org/10.53061/QVLQ9093

A CIP record for this book is available from the British Library

ISBN 978-1-78427-573-0 Pbk
ISBN 978-1-78427-574-7 ePub
ISBN 978-1-78427-575-4 PDF

EU Authorised Representative: Easy Access System Europe – Mustamae tee 50,
10621 Tallinn, Estonia, gpsr.requests@easproject.com

Cover images © Unsplash/Touann Gatouillat Vergos & iStock/xijian

Typeset in Palatino LT Std by S4Carlisle Publishing Services, Chennai, India

For Andy

CONTENTS

ACKNOWLEDGEMENTS

This book came to me at a moment when I was looking for a new big project. I was washing my hands and Boom! – the idea for it erupted in my mind, fully formed like Athena from the head of Zeus. Prior to writing this book, I had created several data management handouts and worksheets to use in lectures and workshops. My big idea was to expand my worksheet offerings to cover the entire data management lifecycle. Basically, I wanted a workbook of activities to pull from for any data management lesson I taught. The result is this book.

Books are never truly written by only one person. While I did the majority of the labor for this book, it would not have been as helpful, accurate, or complete without the assistance of other people.

First, thank you to several people at Caltech Library for help with this book. Thank you Donna Wrublewski for being on board with this big idea and supporting it. I appreciated Catherine Geard and Stephen Davison's support to turn this into a traditionally published book. Big thanks to Tom Morrell for critical feedback on the first rough draft of this book, particularly on the storage exercise which was really terrible before his feedback.

Thank you to Abigail Goben and Dorothea Salo for always being encouraging when I come to them with the idea for a new book, no matter the idea. Abigail was one of the first people I discussed the idea with, which helped cement the concept in my head. Thank you to Rachel Woodbrook for feedback on the accessible spreadsheets exercise.

I greatly appreciated Jamene Brooks-Kieffer and Tina Griffin's feedback on the full draft. Their outside perspective was invaluable for where the exercise instructions were not clear.

Thank you to Erin Wiegand at Pelagic Publishing for advocating to publish this book.

Finally, thank you to my husband, Andy, for being calm and supportive when I had the idea for yet another book (it helped that this book is shorter than my previous two). And thank you to my children, Hank and Wilhelmina, for cuddles, questions, and all of our adventures that keep me grounded.

1

INTRODUCTION

In the autumn of 2019, in her first semester as a faculty member at the University of California-Davis, Kate Laskowski received an email from a trusted colleague about some weird, duplicate values in the data from an article she co-authored in 2016 (Laskowski 2020, 2024). The article was about individual and group behavior in spiders (Laskowski, Montiglio and Pruitt 2016), and Laskowski became concerned that these issues could impact the integrity of the article. She decided to investigate the source of the errors. The final data had been publicly shared in the Dryad data repository, and thankfully she still had the code she used for her analysis. Examining her data and code helped her to determine that the duplicate values were original to the raw data and a not a product of analysis.

Laskowski reached out to her collaborator Jonathan Pruitt, who collected the original data in his laboratory, to determine the source of the errors. Pruitt explained that the duplicate values resulted from a sloppy methodology – his lab measured multiple spiders simultaneously instead of each spider individually, which meant that some spiders were assigned the same values. Pruitt wanted to publish a correction to the article's methodology, but Laskowski decided to first confirm that the data fit this explanation and that nothing else was going on.

Digging into the article's data at a deeper level, Laskowski did not find duplicate values across groups of spiders that might have been measured simultaneously. Instead, she found duplicate values from individual spiders across multiple measurements over time (each spider was measured multiple times in the study). She also found that individual spiders sometimes had two values over time that differed by exact multiples of 100, as well as groups of identical measurements for different spiders. Not trusting the original data provided by Pruitt, Laskowski retracted the article (Laskowski, Montiglio and Pruitt 2020; Marcus, 2020b).

Laskowski then decided to examine the data from two previous articles she co-wrote with Pruitt. She discovered similar duplicated values and decided to retract both articles (Laskowski 2020, 2024). As word of these articles' retractions spread, other of Pruitt's collaborators started to analyze data that Pruitt had collected for them (Marcus 2020a). These investigations resulted in a trail of evidence of data anomalies across multiple of Pruitt's published articles. Journal editors and Pruitt's university soon got involved in trying to determine which articles were suspect and how to handle Pruitt and his research.

In 2023, Pruitt's employer McMaster University released their research misconduct report, finding that the researcher "engaged in data falsification and fabrication in several papers" (McMaster University 2023); he and McMaster had already parted ways the year before (Kincaid 2023; López Lloreda 2023). As of 2024, Pruitt had 18 retracted articles, 12 articles with expressions of concerns, and 4 corrected articles (The Center for Scientific Integrity 2025). Pruitt's scientific career was over, all resulting from a question to Laskowski about their publicly available research data.

As for Laskowski, she has become a firm believer in the transparency of data and code to promote trust in science. She credits her public honesty about the problems with her and Pruitt's data with helping her avoid the taint of Pruitt's scientific misconduct (Laskowski 2024). While the scandal was not an ideal way to start her time as a faculty member, she came out of it with an increased resolve to better document and share her research. In the blog post describing why she retracted her articles with Pruitt, she wrote:

> Each year, I learn about and then attempt to incorporate new protocols to make my science more open and transparent. I am a consummate note-taker, schedule-maker, and photo-taker. I used to upload my data when it was required, but in the past few years, I am now doing it as default. Last year, I made a resolution that I would upload all the R code I used for analyses in my papers. I'd like to take a moment to strongly encourage this as standard practice for all scientists; analysis is such a critical step of any experiment and the more eyes I can get looking over what I did, the better. Science thrives in the light of day! (Laskowski 2020)

1.1   What is Data Management?

This book starts with the story of Kate Laskowski, not because an overwhelming number of researchers are fabricating data, but rather to understand how to be prepared when you want to take a second look at your research. This problem arises frequently in research, such as when someone needs to reproduce or build on the work of a person who has since left the research group, or you need to remember the precise details from an article you published five years ago, or you want to repurpose a figure and its data for a new study. And sometimes, you just need to reexamine your data and notes from earlier in your current project because you forgot specifics of how the data was collected, even if you only collected it last month! In all of these cases, you need to be able to answer the following questions in the affirmative:

- Can you find the data and code?
- Are you sure that is the correct set of files?
- Can you open the files and run them?
- Is everything documented well enough for you to understand the data collection and analysis, even if they were done by someone else?
- Can you actually reproduce the data and analysis?

How well you answer these questions determines the quality of your "data management".

The term "research data management" has become more common in research fields since the 2010's, particularly due to modern funding agency policies around data management and sharing (Government of Canada 2021; White House Office of Science and Technology Policy (OSTP) 2022; EPSRC 2024; European Commission 2024). Data management is commonly encountered when writing a "data management plan" (DMP) or "data management and sharing plan" (DMSP) when applying for a grant. The DMP describes how you are going to organize, document, and share your data during the planned project. This bureaucratic view of data management – as described by funding agencies – has its place, but this book takes a more practical approach in that data management should actually improve your day-to-day experience conducting research.

In this book, "research data management" refers to a set of collective practices and decisions that make it easier for you, your future self, your collaborators, and anyone else working with your data to find, understand, and use that data. These practices cover the entire lifecycle of research data (see next section), from its collection and analysis through sharing and reuse. Data management also applies to all types of research, from science to social science to humanities research; so long as you have data, it needs management. There is no one magical data management practice to solve every data problem. Rather, data management consists of a number of small activities that make dealing with your data a better experience. This means that you can spend more time doing research and less time contending with your files.

1.2   The Research Data Lifecycle

A book about data management would not be complete without mentioning the research data lifecycle. But because the lifecycle is a standard touchstone in data management, everyone has their own version (if you're curious to see the many versions, search for "research data lifecycle" in your favorite online search engine). The version for this book is in Figure 1.1.

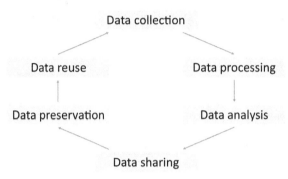

Figure 1.1: A research data lifecycle consisting of six stages (data collection, data processing, data analysis, data sharing, data preservation, and data reuse) with arrows pointing from one stage to the next in a loop, showing how project activities generally flow around the lifecycle.

This research data lifecycle consists of six stages: data collection, data processing, data analysis, data sharing, data preservation, and data reuse. The idea is that a research project usually starts with collecting data, followed by performing other actions on that data such as analysis and sharing, and ending with existing data potentially getting reused in a new project. The lifecycle is a generalization – for example, researchers usually bounce between data collection, processing, and analysis rather than doing them strictly sequentially – but it is a useful tool for understanding how data undergoes changes across the research process.

In this book, the specific stages of the research data lifecycle are less important than the idea that data goes through different stages over the course of its lifetime. Each stage indicates that different actions are taken on that data. And just like data actions change across the lifecycle, so too do data management activities. For example, file organization (Exercise 3.1) and naming conventions (Exercise 3.2) get set up during the data-collection phase of a project, but you only create an archive folder of your important files (Exercise 7.2) at the end of project during the data preservation stage. It can be helpful to think of data actions and data management activities within this lifecycle framework to help you organize and plan when to perform different data management activities.

To align with the research data lifecycle, you should expect to do the data management exercises at the beginning of the book toward the beginning of the research data lifecycle and to do the exercises at the end of the book toward the end of the research data lifecycle. However, there isn't a perfect correlation between the organization of exercises in this book and the phases of the research data lifecycle. For instance, most of the documentation exercises (see Chapter 2: Documentation) should be done during data collection, processing, and analysis, but you will also need to write good documentation during the data sharing phase of the data (see Chapter 6: Data Sharing). The data lifecycle (both the one in this book and data lifecycles generally) is not perfect or absolute, but the hope is that it will still help you plan out your data management practices better.

1.3   Why Do Data Management?

If you conduct research for long enough, you will likely experience a problem stemming from your data. A recent study of the frequency and severity of data management mistakes reported that a quarter (25%) of survey respondents experienced data management mistakes with moderate, high, or very high frequency (Kovacs, Hoekstra and Aczel 2021). Even those who didn't frequently experience data mistakes faced consequences from the mistakes that did happen. When asked about the most severe data management mistake in their laboratory from the past 5 years, 46% of the survey respondents reported that the mistake led to moderate consequences and 22% reported that the mistake led to major or extreme consequences (Kovacs, Hoekstra and Aczel 2021). Clearly, data management errors are happening and sometimes require significant time and effort to fix.

Some of the most common data management mistakes reported by Kovacs included:

- ambiguous naming/defining of data
- version control error
- wrong data processing/analysis
- data coding error
- loss of materials/documentation/data
- data transfer error
- data selection/merging error
- bad or poor documentation (Kovacs, Hoekstra and Aczel 2021).

It's likely that you have experienced at least one of these problems if you've been doing research for a while. It may make you feel better to know that data problems are a common experience for those conducting research. Most people struggle with data management because it is not systematically taught as part of the research process (Tenopir et al. 2016).

The good news is that there is a way to prevent such data problems from happening in the first place through good data management. Data management, done well, means spending less time searching for specific files and understanding what's in them. It means being able to easily pick up a project that's been languishing for 6 months or a coworker's project after they leave the institution. Doing good data management means having access to your data after a natural disaster. Or knowing what rights you have to data from a collaborative research project. None of these things is really about your research but all of them make your research easier.

Good data management doesn't require you to completely rework your research workflows. As mentioned earlier, data management is a combination of many small practices that can be integrated into the work you are already doing in order to make your research more efficient. The ultimate goal is to have your research be the hard part of your work, rather than any data problems taking up valuable research time.

1.4   Using This Book to Manage Your Data Better

Presumably, you picked up this book because you want to better manage your research data. If you have heard of data management and want to do it better, even and especially if you don't know how, this is the book for you. This book will help you implement a number of practical strategies for improving your data management.

This book contains a collection of exercises that will help you to implement data management strategies within your existing research workflows. These exercises – worksheets, checklists, and procedures – will aid you in setting up new data management practices and improving existing data practices, and guide you toward making good decisions about your data. Once these exercises' data practices are implemented, you should feel more comfortable answers the questions in Section 1.1 about finding and using your data.

This book is not a foundational text about data management; for that, see my first book *Data Management for Researchers: Organize, Maintain and Share Your Data for Research Success* (Briney 2015). The exercises do not assume background knowledge of data management, though knowing why data management is important will help you understand why each exercise is valuable and where it fits within the research data lifecycle. To aid with this, each exercise has an introduction to help you contextualize the problem being solved. If, in the course of working through an exercise, you want to learn more about a specific data management topic, you are encouraged to refer to one of my previous books, *Data Management for Researchers*, for further background. Table 1.1 lays out how the general topics in this book correspond to the same topics in *Data Management for Researchers*, chapter by chapter.

The book loosely aligns with the data lifecycle in Figure 1.1 but does not provide a complete set of exercises for everything under the umbrella of data management. Instead, the collection centers on activities that are structured, reproducible, and apply to many researchers. The strength and weakness of research data management is that many of its practices are customizable to individual research workflows. Each exercise in this book, therefore, is built on best principles while allowing for customization to suit your particular needs.

Table 1.1: Topical correspondence between *The Data Management Workbook* and *Data Management for Researchers*.

Topic	*The Data Management Workbook*	*Data Management for Researchers*
Documentation	Chapter 2	Chapter 4
File organization and naming	Chapter 3	Chapter 5
Data storage	Chapter 4	Chapter 8
Data management	Chapter 5	Chapter 3
Data sharing	Chapter 6	Chapter 10
Project wrap-up	Chapter 7	Chapter 9

You may go through exercises in this book collectively or individually, as you choose. Do note that some exercises require completing one or more other exercises in the book; this is to help focus each exercise on a single data management task. If the whole collection of exercises feels overwhelming, start with one exercise and spend time making sure that practice is routine before tackling another exercise, and so on.

For the exercises that have been formatted as worksheets, you are encouraged to write your answers in the space provided. These exercises are also accompanied by examples to help you understand one possible answer to each prompt. Most of the examples come from my own research experiences in chemistry and librarianship, though a few are made up entirely. The hope is that the examples provide enough context for reimagining a solution to fit within your specific research workflows.

Hopefully, by the end of the book you will feel more comfortable with implementing data management into your research workflows. You don't have to be perfect; small data management actions can make a big difference in saving you time and energy later. Do your best. I will consider this book to be a success if these exercises help make your research a better experience. Good luck!

2

DOCUMENTATION

I spent most of a year of graduate school for chemistry reproducing someone else's research. The work served as the foundation for the experiments I actually wanted to conduct, so I needed to deeply understand it. While I had the research data and lab notebooks from the person who did the original study, the documentation wasn't good enough to understand the research at the level I needed. The result was a lot of frustration and my first major understanding of the power of documentation to make or break research.

If you have ever spent time looking at an older data file and trying to determine what it contains, how you collected it, or if it is the version of the file that you are looking for, you too have experienced how important documentation is to data management. Documentation has sometimes been called "a love letter to your future self" as it helps you remember important details about your research data. We are liable to forget such details over time, so documentation is important for providing the necessary context to enable data use and reuse by both ourselves and others.

The great thing about research documentation is that it's not limited to a laboratory or research notebook, though notebooks are still very important. While we will review the quality of your notes, this chapter encourages you to try out other types of documentation such as a project information file, a data dictionary, or a template. These other forms of documentation are not universal but fill important niches for recording the context of research data, ones that make it easier to share and reuse that data. So try out this chapter's exercises and learn how to make documentation work better in your research.

Exercise 2.1: Evaluate a Laboratory/Research Notebook

The laboratory or research notebook is a fundamental documentation method for many researchers. But for how ubiquitous the research notebook is, its documentation can sometimes be insufficient. The ideal research notebook allows someone with similar training as you to be able to follow everything you did in your research. Imagine winning the lottery and quitting your job – would your colleague be able to find your notes and restart the research where you left off? If you answer no to this question, find ways to improve your notetaking. Even if you answer yes, there are probably still things you need to clarify for someone else to understand. Notetaking is an art that can always be refined.

This exercise has you review an old entry in your laboratory/research notebook to evaluate if your documentation is sufficient for reproducing your work. You will need an older laboratory/research notebook entry, ideally from 6–12 months ago (long enough in the past to have forgotten some details but not everything), to do this exercise. Once you have the entry, read through it to try to understand what you did on that day. Answer the exercise questions to evaluate the entry and identify any note keeping improvements to make. As much as possible, answer the questions using the information in the notebook entry rather than your memory of it.

1. Date of research notebook entry being evaluated:

   Example: 2024-09-17

2. Read the entry and summarize the work you did on that date:

   Example: On this date, I searched university websites for university data policy. I got through A–J in the alphabetic list of Carnegie R2 universities and noted when the university had a data policy and its URL. I also downloaded a copy of each policy that I found in preparation for doing a textual analysis of the policies.

3. How easy was it to understand the work you did from your notes?

   Example: It was somewhat easy to understand the work I did, though my hand-writing could be a bit tidier. It was pretty clear as to the parameters of the work done on that day, specifically which group of university websites I searched.

4. Could you reproduce your work based on the information in your notes? If not, what extra information do you need?

   Example: Sort of. I had to look elsewhere in my notes for the terms I used for searching and I think I used Google Search, but it's not exactly clear. Thankfully, the front of my notebook had an outline of my data management plan, so I know where I saved the downloaded policy files.

5. What worked well with your note keeping?

   Example: I did very well laying out my daily page so it's distinct from other days' entries and the boundaries of the work I did were clear.

6. What should you improve about your note keeping?

    Example: I need to improve my cross-referencing. For example, I wrote the search terms in another notebook entry but I didn't explicitly note where these were written in this day's entry, meaning I had to visual scan through my notes to find them.

7. List one change you will implement to take better research notes:

    Example: I will implement cross-referencing, specifically by noting the date of the relevant notebook entry in (YYYY-MM-DD) format, so that I don't have to repeat information and can easily find it later.

Exercise 2.2: Document Project Information

Data files living on a computer often need extra documentation in order to understand what research they correspond to, especially if your main research notes are stored separately from your data, such as in a paper notebook. In particular, it is useful to record basic project information and store this documentation in the top-level folder for research project to give context for the project's files. Imagine coming back to these files in the future – the project description document should help reorient you to the project. This exercise walks you through the key project information to record.

To complete this exercise, pick a research project and answer the questions. Copy all of the written information into a TXT file (almost all word processing software can open a TXT file) and save it with a descriptive file name. Store this file in the top-level of the project folder on your computer, alongside the project files. The same information can also be written in the front of a physical research notebook. Repeat this process for each new research project.

This exercise was adapted from the "Project Close Out Checklist" (Briney 2020c).

1. What project is being documented?

   Example: The Data Doubles project.

2. Write a brief project description:

   Example: The Data Doubles project was a 4-year, IMLS-funded research project examining student perceptions of privacy in library learning analytics.

3. What is the time period the project was done over?

    Example: The project was conducted between summer 2018 and summer 2022.

4. Who worked on the project?

    Example: Eight researchers from eight different institutions worked on the project, including: KMLJ, AA, KB, AG, MP, MR, DS, and MS.

5. Where are the data, code, and other files are stored?

    Example: Research files are stored on Google Drive, with the exception that participant data is stored in IU-hosted Box. Survey data is also in Qualtrics. Code is on GitHub. The shared literature library is in Zotero. All grant documents are also stored in Google Drive.

6. How and where is the project documented?

   Example: Documentation for the project is in Google Drive. Notes on team decisions from meetings are in the DataDoubles/Meetings folder. Notes on data are in the DataDoubles/Research folder.

7. How are files organized? Are any naming conventions used and, if so, what are they (see Chapter 3)?

   Example: All data is in the DataDoubles/Data folder, with subfolders labelled by interview theme code. Each site has its own folder within the project folder for individual site files. Interview data files are named with: interview theme, site, interview ID, interview date, and data type/analysis stage (e.g. "PRO_BL03_20180222_Audio.mp3" and "AWA_MK01_20180222_Notes.pdf"). Please see the living data management plan for a complete set of codes and more details.

8. What else does someone need to know to understand these files?

   Example: Additional documentation on the project and the project's public research files are available on OSF.

9. Record all of this information in a TXT file, give it a descriptive name, and save it in the top-level folder with your project files.

Exercise 2.3: Create a Data Dictionary

Spreadsheets are one of the most common file types used in research, but they are not always formatted for data reuse. Ideally, a spreadsheet is formatted with a row of variable names at the top, followed by rows of data going down (see Exercise 6.1: Make a spreadsheet more accessible and reusable for further spreadsheet formatting tips). This formatting makes it easy for data to be used in any data analysis software, but limits how that data can be documented as you cannot add all of the necessary contextual information to the file itself. For this reason, it's useful to create a separate data dictionary to describe the spreadsheet so that others can interpret the data. A data dictionary defines each variable in a spreadsheet, its units, and other variable information necessary for interpreting a spreadsheet's data

This exercise walks you through the key information you should record for each variable in the spreadsheet, adding up to a complete dictionary to accompany the spreadsheet file. To complete the exercise, identify a spreadsheet, pick one variable, and record its information in a column of the table. Repeat this process for the remaining variables in the spreadsheet. (Note: there is likely not enough space in this exercise to cover all variables in your spreadsheet. Start the data dictionary using the supplied structure and complete it digitally using the same information for all variables.) Copy the complete set of variable information into a text document or separate spreadsheet and save it alongside the spreadsheet. It is useful to save the data dictionary with the same root file-name as its data file but appending "_dictionary" on the end of the file name; for example, the data dictionary for the file "myData.xlsx" would be "myData_dictionary.txt."

This exercise was adapted from "Leveling Up Data Management" (Briney 2023).

1. Spreadsheet documented by this data dictionary:

   Example: "2025_NationalParks_WaterQuality.xlsx" which contains water testing information from several U.S. national park sites.

2. Record the following information for each variable in the spreadsheet:

Variable Information	Example Variable	Variable 1	Variable 2	Variable 3	Variable 4
Variable name	*site*				
Variable description	*Two-letter abbreviation describing the name of the overall site where the sample was collected.*				
Variable units	*N/A*				

Variable Information	Example Variable	Variable 1	Variable 2	Variable 3	Variable 4
Relationship to other variables (as applicable)	*Partner to variable "sampleNum," which together define the sample ID (site name + sample number at that site). Related to variables "latitude" and "longitude," which record exact coordinate location and are more specific than the larger site code.*				
Variable coding values and meanings (as applicable)	*Coding values and meanings: BL = Badlands NP; DV = Death Valley NP; GT = Grand Teton NP; JT = Joshua Tree NP; ZN = Zion NP*				
Known issues with the data (as applicable)	*Some Badlands samples were collected outside of the park boundaries; see latitude and longitude variables for specific locations.*				
Anything else to know about the data?	*Older data (pre-2013) used one-letter abbreviations for site code but this was updated for clarity and ease of identification.*				

3. Complete the exercise for all spreadsheet variables and document the information in a text file or another spreadsheet. Give the data dictionary the same filename as the spreadsheet file but with "_dictionary" added onto the end of the file name, and save it in the same folder as the spreadsheet.

Exercise 2.4: Build a Documentation Template

Surgeons use checklists at the beginning and end of surgery to make sure that they are following the proper safety protocols and to reduce the possibility of errors during surgery. In a similar way, researchers can use templates to reduce the possibility of documentation errors during routine data collection. A template is a standard list of information that should be recorded every time a specific type of data is collected. Using a template will make your documentation more consistent and ensure that you don't forget any important information. (And while this exercise is about building a template for documentation, you can also use templates for routine data analysis or even to help you implement a new data management task!)

This exercise prompts you to identify a specific type of data and then brainstorm information that needs to be documented whenever you collect that data. Note: If your data is captured automatically, your list should include experimental parameters outside of your data; if you are recording data by hand, your list should also include all of your data variables as well as the experimental parameters. You can format this list either as a paper worksheet or as an electronic spreadsheet. Once you have your template, be sure to use it every time you collect that specific type of data so that you have complete documentation.

1. Identify a specific type of data, e.g. from a regular experiment or observation, that you frequently collect.

    Example: UV-Vis laser spectroscopy of a chemical reaction.

2. Brainstorm a list of all of the information you should record when you collect that specific type of data.

    Example: laser power, laser wavelength, lab humidity, solvent, solute 1, solute 2, solute 1 concentration, solute 2 concentration. The actual spectrum is collected by the computer.

3. Arrange the list in the most logical chronological order, representing how you would collect this information within your research workflow. Put the date at the top of the list.

    Example: today's date, lab humidity, laser wavelength, laser power, solute 1, solute 1 concentration, solute 2, solute 2 concentration, solvent.

4. Specify units or formatting for items on the list, as applicable.

    Example: today's date (YYYY-MM-DD), lab relative humidity (%), laser wavelength (nm), laser power (mW), solute 1, solute 1 concentration (g/mL), solute 2, solute 2 concentration (g/mL), solvent.

5. Print the list out as a paper worksheet or format it as an electronic spreadsheet to make it usable as a template. Fill in the template every time you collect that type of data. Update the template as needed over time.

3

FILE ORGANIZATION AND NAMING

One of my greatest successes in data management happened when I convinced my research colleagues to use a standard file organization and file naming convention in our shared research. The project in question involved seven different researchers conducting over a dozen interviews at each of eight different sites. Each interview corresponded to one of five possible interview themes and resulted in: an audio file, interviewer notes, a case summary, a transcript, and a coded transcript. In short, there were a lot of files that correlated in many different ways. To combat potential data management chaos, the project used a rigid folder system and labelled all interview files with: interview theme, interview site, interview number, date, and content type (Briney et al. 2022). This made it very easy to identify which file was which, as well as visually scan through the files to see what was missing – a useful practice during the data collection phase of the project. By the end of the project, I had convinced the project's principal investigator that all of his future projects needed such robust file organization and naming to be successful.

Good file organization and naming are foundational data management practices, as they help you find files quickly when you need them. They also prevent future guessing about what is in a file or if it's the right version – the file name will tell you! To set up file organization and naming conventions, this chapter offers two exercises: a card-sorting process for brainstorming a file organization system; and a worksheet for creating a file naming convention for a group of related files. If you do only one exercise in this entire book, I encourage you to try the file naming convention exercise; it's a basic practice that will save you lots of future time and heartache once you've set up a good system. However, this chapter's two exercises work best when done together, for example by setting up a file naming convention for your organized folders.

Exercise 3.1: Set up a File Organization System

Implementing a file organization system is the first step toward creating order for your digital research data. Well-organized files make it easier to find the data you need without spending lots of time searching your computer. Every researcher organizes their files slightly differently, but the actual organizational system is less important than making sure you use it. Your organizational system will be successful so long as your files have a logical place that they should be stored and they actually get stored in that place.

This exercise prompts you to brainstorm organizational groupings and hierarchies to come up with an order for managing your research data. If you are conducting collaborative research, do this exercise with your research partners so that everyone can agree on the organization scheme. This is a card-sorting exercise, meaning you will need a stack of note cards or sticky notes to do this activity – ideally in three different colors. Follow the instructions to label cards and move them around until you develop your organizational system. An example is provided in Figure 3.1. There is no one correct way to do this so feel free to play around, add or remove cards, and move cards however you want! Once you put your new organizational system into place, be sure to always put your files where they're supposed to go.

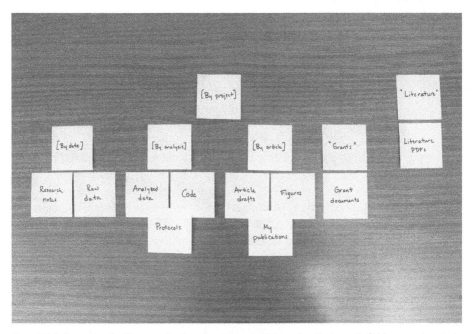

Figure 3.1: Example of a file organizational system on sticky notes. Systematic folders are denoted with text in brackets, single folders have text in quotes, and sticky notes with regular text are file types.

1. Brainstorm the types of files you will create during a project. Consider the following types of files, though you can add, remove, or modify file types as needed:
   - Raw data
   - Analyzed data
   - Code
   - Protocols
   - Article drafts
   - Figures
   - My publications
   - Literature PDFs
   - Grant documents
   - Research notes

2. Write your brainstormed file-type categories on individual cards of the same color. Take a stack of note cards or sticky notes in the first color and write your brainstormed file types on one card each.

3. Start grouping file types together in a way that makes sense for your research. Move cards around and group together file-type cards that you want to store together. Files that will be stored near each other in a folder hierarchy, but not together, should be placed near each other while file types expected to be stored completely separately should be away from other cards.

4. Create folder cards to hold your grouped file-type cards (but don't place the folder cards yet). You can name these cards however you want, but you should represent each card as one of two different folder types:
   - Cards in the second color represent a unique folder that you only have one of. Label this type of folder with the folder name in quotations (e.g. "Literature" or "My publications").
   - Cards in the third color represent multiple folders organized in a systematic way, such as for folders organized by date or by project. Use only one card to represent the organizational pattern that will be repeated. Label this type of folder card with the organizational system in square brackets (e.g. [By date] or [By project]). Note: folders organized by date should be labelled using the convention YYYYMMDD or YYYY-MM-DD to facilitate chronological sorting.

5. Place your new folder cards relative to the file-type cards they should contain. Move file-type cards underneath the new folder cards to show the hierarchy of how a file type will be saved in a specific folder or group of folders. Systematic group folders (cards in the third color) only need to be represented once in the card sorting, as they are assumed to represent multiple folders on a computer.

6. Make final adjustments. Make copies of any type of card and add folder levels, as needed. Adjust placement and hierarchies until you are happy with the organizational system you developed.

7. Record your organizational system in your research notebook and/or the project description file (see Exercise 2.2: Document project information).

Exercise 3.2: Create a File Naming Convention

If I only have time to teach students one data management strategy, I almost always choose to teach file naming conventions. Consistent file naming is a basic way to add order to your files, which has enormous benefit. Not only do rich and descriptive file names make it easier to search for files (both digitally and visually), but good names also help you understand what your files contain at a glance and tell related files apart. For example, the name "data.tif" doesn't tell you what's in this file whereas "JWST_2024-01-26_MIRI_ IC2163-NGC2207.tif" gives you some clues about an image taken by the James Webb Space Telescope (see Figure 3.2), even if you don't know the specifics of the file naming convention. It only takes a few minutes to come up with a file naming convention and a few seconds to properly name a file, but good file naming can save you hours of time later when searching for specific files.

On a side note: this exercise introduces you to the concept of "metadata," which can be summarized as "data about your data". For example, metadata about a dataset can include: the data's title, authors, file size, last updated date, file format, version number, etc. Basically, it's the information that describes the

Figure 3.2: Mid-infrared image of galaxies IC 2163 and NGC 2207 taken by the James Webb Space Telescope (NASA et al. 2024). Used under public domain.

dataset but isn't actually part of the dataset. In the case of this file naming exercise, the metadata will be descriptive information about what the file contains and/or the parameters of the data collection. The idea of metadata will appear again in Exercise 6.6: Share data, which asks you to decide on the necessary citation information (like authors and dataset title) for a shared dataset.

This exercise guides you through the process of creating a file naming convention for a group of related files. Determine which files the naming convention will cover, then follow the subsequent instructions. An example for microscopy data is provided. Consider working though this exercise multiple times, as different file groups and folders usually require different naming conventions. When finished, record your naming convention in your project description file (see Exercise 2.2: Document project information) and write it in your research notebook.

This exercise is based on the "File Naming Convention Worksheet" (Briney 2020b). The worksheet won a Significant Achievement Award in the 2023 DataWorks! Prize competition from The Federation of American Societies for Experimental Biology (FASEB) and the U.S. National Institutes of Health (NIH) (NIH Office of Data Science Strategy 2024).

1. What group of files will this naming convention cover?

    You can use different conventions for different file sets.

    Example: This convention will apply to all of my microscopy files, from raw image through processed image.

2. What information (metadata) is important about these files and makes each file distinct?

   Ideally, pick three pieces of metadata; use no more than five. This metadata should be enough for you to visually scan the file names and easily understand what's in each one.

   Example: For my images, I want to know date, sample ID, and image number for that sample on that date.

3. Do you need to abbreviate any of the metadata or encode it?

   If any of the metadata from step 2 is described by lots of text, decide what shortened information to keep. If any of the metadata from step 2 has regular categories, standardize the categories and/or replace them with 2- or 3-letter codes; be sure to document these codes.

   Example: Sample ID will use a code made up of: a 2-letter project abbreviation (project 1 = P1, project 2 = P2); a 3-letter species abbreviation (mouse = "MUS," fruit fly = "DRS"); and 3-digit sample ID (assigned in my notebook).

4. What is the order for the metadata in the file name?

Think about how you want to sort and search for your files to decide what metadata should appear at the beginning of the file name. If date is important, use ISO 8601-formatted dates (YYYYMMDD or YYYY-MM-DD) at the beginning of the file/folder names so dates sort chronologically.

Example: My sample ID is most important so I will list it first, followed by date, then image number.

5. What characters will you use to separate each piece of metadata in the file name?

Many computer systems cannot handle spaces in file names. To make file names both computer- and human-readable, use dashes (-), underscores (_), and/or capitalize the first letter of each word in the file names. A good convention is to use underscores to separate unrelated pieces of metadata and dashes to separate related pieces of metadata for parsing and readability.

Example: I will use underscores to separate metadata and dashes between parts of my sample ID.

6. Will you need to track different versions of each file?

You can track versions of a file by appending version information to end of the file name. Consider using a version number (e.g. "v01", "v02"), the version date (use ISO 8601 format: YYYYMMDD or YYYY-MM-DD), or format (e.g. "_audio", "_transcript"), depending on your needs.

Example: As each image goes through my analysis workflow, I will append the version type to the end of the file name (e.g. "_raw," "_processed," and "_composite").

7. Write down your naming convention pattern.

Make sure the convention only uses alphanumeric characters, dashes, and underscores. Ideally, file names will be 32 characters or less.

Example: My file naming convention is "SA-MPL-EID_YYYYMMDD_###_status.tif" Examples are "P1-MUS-023_20200229_051_raw.tif" and "P2-DRS-285_20191031_062_composite.tif."

8. Document this convention in the project information file and record it in your laboratory/research notebook.

4

DATA STORAGE AND SECURITY

Since I moved to California, I have become very familiar with earthquakes and wildfires. I follow emergency preparedness guidelines and always have a bag packed in case I need to leave my home quickly (in the event of a wildfire), as well as a storage tub full of supplies in my house in case I'm stuck at home for multiple days without water or power (in case of an earthquake). In a similar way, I disaster-proof the storage of my data (in the event that my home or office is destroyed by either type of disaster) by keeping a copy of all my data in a geographically distinct location. Not only does this protect against natural disasters, it also safeguards against more frequent scenarios such as hard-drive failure or laptop theft. For instance, I had a friend in graduate school who was halfway through writing their PhD thesis when their laptop was stolen; they ended up having to rewrite the entire document from scratch because they didn't have a backup copy. Unfortunately, such stories of data loss are common.

This chapter helps prepare you for the scenario of your main research computer being gone instantaneously, either because it was stolen, died, was destroyed, etc. You need to have a backup for your data and it's recommended that your backup is far away from where your computer lives. For those with Big Data, this can be especially challenging as storage on the scale of terabytes and petabytes quickly becomes expensive; do the best you can and focus on backing up your most important data. Beyond making sure you have a backup system in place, this chapter will also help you test your backup system (it's much better to learn how to do this before disaster strikes and you're stressing about trying to recover data) and make sure your data is stored safely behind a strong passphrase.

Exercise 4.1: Pick Storage and Backup Systems

Research data needs to be stored and backed up reliably so that important data is not lost. But choosing storage can be a challenge, as there are many commercially available options and institutions don't always offer uniform services for storage and backup. Big data compounds this problem, especially as costs for large storage scale. If you are struggling with picking the right storage and backup systems, it can be helpful to list out all possible options, note their costs and benefits, and select the right combination that best meets your needs.

For the purposes of this exercise, we'll classify storage into two types: a system where you regularly interact with and manipulate your files (we'll call this "working storage"); and a system that contains a copy of your data, just in case, that you don't interact with frequently (we'll call this a "backup"). It is possible to have more than one working storage area, but be aware that this setup can get messy if you are moving files between working storage areas and trying to ensure you have the correct versions of these files backed up. For backups, you need at least one. To make things easier, this exercise assumes you will have one working storage area and one, maybe two, backups.

This exercise prompts you to examine the working storage and backup systems available to you before determining which is the best set of options for your data. Answer the questions and then fill out the table of information about each possible working storage and backup system. Evaluate all of the options in the table using the listed criteria to select a working storage location and backup system and, as needed, an alternative backup. There usually isn't a perfect solution for working storage and backup options, so pick what will best meet your needs.

1. What is the estimated total data storage you will need over the next five years?

    Example: I estimate that I will generate 100 GB of data over the next five years of my project.

2. Does your data require meeting any specific security standards? If so, what level of security?

   Example: My data will include some human subjects data, so my storage systems must have restrictions on access, but it's not medical data so they don't have to be HIPAA compliant.

3. What are your top one or two concerns about your storage and backup, as identified in Table 4.1?

   Table 4.1: List of common concerns around storage and backup systems and how to mitigate those concerns.

Concern	How to mitigate concern
Storage size for large data	Only backup key data
Budget limitations	Use systems provided by your institution
Limited technical knowledge to run storage and backup	Use systems provided by your institution
Forgetting to back up files	Use automated backups
Losing data due to viruses/ransomware	Do not use synchronized backups (e.g. where a file is updated across all devices in real time)
The data needs to have security and controls over who can access it	Discuss security options with local IT
Potential for device failure (e.g. the hard drive might die)	Store and back up your files on different types of storage devices
Potential to lose data due to environmental disaster (e.g. fire, hurricane, earthquake, etc.)	Keep a copy of your data in a different geographical location
Project leader needs to retain files when group members leave	Use systems that have administrative privileges that can manage accounts for groups

Example: I live in an earthquake zone, so having storage and backup that can withstand an environmental disaster is a top concern.

4. What working storage and backup systems are available to you, such as through your institution, workplace, or elsewhere?

   Example: I have the following systems available to me: my laptop, a Time Machine backup, a departmental server, and institution-provided Box and Google Drive.

5. Fill out the information in the table for each working storage and back-up system you are considering.

System Attributes	*Example System*	System 1	System 2	System 3
System name	*Departmental server*			
Is it working storage or backup?	*Working storage*			

System Attributes	*Example System*	System 1	System 2	System 3
What is the cost?	*No cost for 10GB and under. Cost is $5 per 10 GB per year after that.*			
What is the hardware type?	*Server, exact hardware type unknown.*			
Is the system backed up?	*No backup.*			
For backup systems, is backup automatic?	*N/A*			
What level of security does the system provide?	*Storage is password protected.*			
Is the system local or remote?	*System is local.*			

System Attributes	*Example System*	System 1	System 2	System 3
Is there a limit to storage capacity?	*Storage limit is 500GB per research group.*			
Who manages the system?	*Departmental IT manages the server.*			
Is it easy or difficult to use?	*Very easy to use once set up.*			
Does the system help mitigate your top concern(s) from step 3?	*No, as the server is in the same location as the laboratory.*			

6. Optimize your working storage and backups based on the following considerations:
   a. You need a working storage system that:
      - will hold all of your data files;
      - meets your needed level of security;
      - allows you to easily interact with and manipulate your files.
   b. You need one backup that:
      - will hold all of your data files;
      - meets your needed level of security;
      - is reliable/managed by someone you trust;
      - is easy to use;
      - backs up automatically.

c. At least one backup should be in a different location than your main storage system for disaster resiliency. Your backup(s) should also help mitigate your top storage and backup concern(s) from exercise step 3. If the first backup does not succeed in addressing these issues, choose a second backup that:
- will hold all of your files;
- meets your needed level of security;
- is reliable/managed by someone you trust;
- addresses your storage and backup concern(s) from step 3.

7. Pick your storage and backup systems and write them down.

Example: My primary storage will be my computer with added security restrictions. I will use Time Machine as my first automatic backup and the institutional Box, which is controlled access, as my second backup because that storage is in a different geographic area.

Exercise 4.2: Test Your Backup

Backups are super important for your data, so it's always good to test that your backups are still working. Nothing is worse than losing your data from your working storage and then realizing that your backup isn't functioning. Beyond checking that your backup is functioning, it's also good to know how to recover your files so that you don't have to learn this for the first time while panicking about lost data. So take a few minutes to test your backup now so as to save heartache later.

This short exercise walks you through getting a file from your backup to test that it is working and to learn how the data-recovery process works. To complete the exercise, pick a backup system and a file to recover and work through the steps. The hard part of this exercise is finding instructions for file recovery and actually recovering the file, which vary by backup system.

1. Identify where your data is backed up.
2. Find instructions for recovering data from your backup system.
3. Pick a data file from your computer.
4. Follow the instructions you located in step 2 to get a copy of the data file from step 3 out of your backup system.
5. If this process didn't work, fix your backup system. If this process did work, congrats your backup is working and you know how to recover your files!

Exercise 4.3: Create a Strong Passphrase

Passwords are a standard security layer for many systems, though they can be easy to crack. Current recommendations are to use a trusted password manager, multi-factor authentication (MFA), and long passwords (NIST 2025). A long password, of at least 15 characters, that is just made up of letters is actually more secure than an 8-character password that uses a mix of character types (NIST 2025). To make it even easier to remember a long password, security experts recommend creating a "passphrase". Passphrases are longer passwords made up of multiple random words strung together (Electronic Frontier Foundation 2016a; Hearn 2016). Passphrases' extra length make them easier to remember but much more difficult to computationally hack, provided that the words in the phrase are truly random and not a quote or a sentence known to the public.

Personally, I use a password manager locked behind a passphrase and MFA, and I let my password manager create the rest of my passwords as mixes of 15+ random characters. These random passwords are very secure but literally impossible to remember. It's much easier for me to remember the passphrase for my password manager and let the password manager do the rest. Plus, I know the passphrase is a strong protection for the rest of my passwords due to its length.

(If you don't use a password manager, I encourage you to try one out. They solve the problem of trying to remember every password or passphrase you create, especially as you are supposed to have a unique password for every system and website. The password manager will remember all of your passwords and can even come up with long, complex passwords to use in various systems. Lists of the best password managers are readily available on the internet.)

This exercise walks you through one method for generating a passphrase: creating and memorizing a collection of random words. Note that a six-word passphrase will be more secure than a four-word passphrase, however some systems have password character limits. Also, do not use the example pass-phrase for your own, as it is less secure for having been published. Even if you only use a passphrase for your password manager, it's still a good security strategy to know about.

1. Write 4–6 random words. You can do this by coming up with random words on your own, flipping through a dictionary, or using a "diceware" passphrase word list (Electronic Frontier Foundation 2016a, 2016b).

   Example: funeral loom dour burning

2. To help you remember the passphrase, create a sentence using those words.

   Example: She held a <u>funeral</u> for her weaving <u>loom</u>, making her <u>dour</u> as the world was <u>burning</u>.

3. If you want more security than just these words, you can add capitalizations, punctuation, and/or random characters to the passphrase, but it is not required.

   Example: Funeral loom! Dour burning.

4. Write down your final passphrase and record it in your password manager, if you use one.

   Example: Funeral loom! Dour burning.

5

DATA MANAGEMENT PLANNING AND POLICIES

Break ups are difficult. Whether you are breaking up with your significant other or your favorite musical group is breaking up, it's often messy and emotional. Unfortunately, research break ups can be difficult too, despite the fact that planned break ups occur frequently as students and trainees leave research groups. Unplanned departures can be even worse. I was once a part of a large, multi-year research project where: one collaborator unexpectedly left after a year; two new researchers came on to replace them; and yet another collaborator (it was me) took maternity leave, came back to the project briefly, and then took another break to move halfway across the country (that was a very busy year). So how did our project handle all of the comings and goings? By being clear about what was happening and what was expected of all collaborators. The team had robust data management plans, which allowed my research assistant to continue some of my work when I was on maternity leave, and had long conversations about rights and access to our collected data. Now that the project has ended, all of us collaborators have a shared understanding of how to handle any future analysis and publication of project data in a manner that respects the others' claim to that data because we talked about this beforehand. All of the team's conversations about data handling and permissions, and the documentation that spawned from these conversations, cut to the heart of data management as they focused on making good decisions about our data.

While this entire workbook covers data management activities, this chapter takes a step back to document data management decisions that should be made. Here you will find assistance with data management plans (DMPs), but you might be surprised to see guidance about two different types of plans. The first is likely what you expect: a data management plan written for a funding application. The second, a living data management plan, is actually more useful day-to-day because it spells out how the data will actually be handled during the project, with the document being updated as needed to reflect actual practices. I find this second type of plan to be highly valuable, particularly for collaborative research where many people access the data. For researchers who are not the project lead, I also encourage you to work through the exercise on data stewardship to better understand what future rights and responsibilities you have over your data; it's better to ask these questions early instead of having a misunderstanding about data you don't actually have permission to use. No matter the exercises, this chapter shows the benefit of discussing data management with those you work with to develop shared understandings of data expectations.

Exercise 5.1: Write a Data Management Plan for a Funding Application

One of the most frequent ways researchers encounter the concept of "data management" is when applying for funding and writing a data management plan (DMP) or data management and sharing plan (DMSP) for a grant application (note: the term DMP has traditionally been used but many U.S. federal funding agencies are now calling them DMSPs). A narrative DMP (as opposed to a form-based or machine-actionable DMP) can feel like a bureaucratic hurdle, but these documents have value in helping funders ensure compliance with data sharing mandates. As they are an important part of a grant application, here is an exercise to make sure that you have all of the basic information in your plan.

This exercise covers the major components of a DMP, particularly in the U.S. context. To do the exercise, compare the contents of your DMP to the items on the list to ensure your plan contains the necessary basic information. Note that content on this list may differ from DMP requirements from individual funding agencies and their programs; always refer to specific requirements in the call for proposals when writing a DMP/DMSP for a grant application. While this exercise won't be perfect for every grant application, ensuring your document covers everything on this list will give you a solid starting point for your DMP.

This exercise was adapted from the "Data Management Plan (DMP) Checklist" (Briney 2020a) and incorporates a few categories from the Federation of American Societies for Experimental Biology (FASEB) "DataWorks! Data Management Plan Challenge Evaluation Rubric" (FASEB 2021).

Describe the data to be collected:
- ☐ Briefly describe the data.
- ☐ List the anticipated file formats.
- ☐ Estimate the size of the data to be generated over the duration of the grant.

Outline data documentation and standards:
- ☐ Briefly describe how the data will be documented.
- ☐ Identify any metadata standards or standard formats that will be used.

Explain necessary tools:
- ☐ Identify any commercial software or tools that are needed to access and use the data.
- ☐ Describe any custom-built software or tools that are needed to access and use the data.

Identify privacy and intellectual property concerns:

- ☐ Declare whether the data will lead to a patent or other intellectual property claim.
  - ☐ If yes, identify how this will impact the management of the data.
- ☐ Declare any privacy or confidentiality laws, policies, or regulations that apply to the data.
  - ☐ If privacy or confidentiality rules apply, describe any security measures that will be put in place to protect the data.
  - ☐ If privacy or confidentiality rules apply, identify who will be responsible for data security.

Describe the data that will be shared:

- ☐ Describe any limitations or controls on data sharing due to privacy, intellectual property, etc.
- ☐ Describe what portion of the collected data will be shared.
- ☐ Identify the intended file formats for the shared data.
  - ☐ If the data is in a proprietary format, describe how it might be made more accessible (see Exercise 7.1: Convert data file types)
- ☐ Estimate the size of data to be shared.
- ☐ Describe other relevant information (e.g. survey instruments, code, etc.) that needs to be shared for reproducibility.

Describe how the data will be shared:

- ☐ Identify a data repository (or repositories) where the data will be shared (see Exercise 6.4: Pick a data repository for sharing). Note if the repository will provide a DOI or other permanent identifier upon deposit.
- ☐ Document the anticipated license under which the data will be shared (see Exercise 6.5: Select a license for your data).
- ☐ Describe when the data will be shared.
- ☐ Estimate how long the shared data will be available.

Determine responsibility:

- ☐ Identify who is responsible for ensuring compliance with the DMP.
- ☐ Document how long data is expected to be retained after the completion of the grant.

Exercise 5.2: Write a Living Data Management Plan

Many researchers are aware of the two-page narrative data management plan (DMP) for a grant application, but you may not be aware of a more everyday type of DMP: a living DMP. This document describes how data will be actively managed during a project and should be updated whenever necessary to reflect current data practices. A living DMP is a useful touchstone for understanding where data lives, how it's labelled, how it moves through the research process, and who will oversee the data management. This type of document is particularly useful for collaborative research, where many researchers will create, access, and analyze the same data files, necessitating robust data management practices to avoid file chaos.

This exercise guides you through the process of creating a living DMP for your research. To complete the exercise, pick a project and answer the questions to build your living DMP. This DMP should be reviewed at major project milestones and revised any time practices change. If you are doing collaborative research, work through this exercise jointly with your colleagues to determine shared conventions or at least get feedback on the draft DMP from your colleagues. Once the DMP is complete, copy the information to a text file and save it in a top-level folder for your research project with a descriptive file name.

1.   Write a short summary of the project this DMP is for:

   Example: This project uses mass spectrometry to identify isotopic composition of soil samples.

2. Where will data be stored? How will data be backed up? (See Exercise 4.1: Pick storage and backup systems.)

Example: The data is generated on the mass spectrometer then copied to a shared lab server. The server is backed up by departmental IT.

3. How will you document your research? Where will your research notes be stored?

Example: Data collection and analysis is primarily documented in a laboratory notebook, organized by date. README.txt files add documentation to the digital files as needed.

4. How will your data be organized? (See Exercise 3.1: Set up a file organization system.)

Example: Each researcher has their own folder on the shared server. Data within my folder is organized in folders by sample site with subfolders labeled by sample ID. Sample ID consists of: two-letter sample site code, three-digit sample number, and date of sample collection formatted as YYYYMMDD (e.g. "MA006-20230901" and "CB012-20100512").

5. What naming convention(s) will you use for your data? (See Exercise 3.2: Create a file naming convention.)

   Example: Files will be named with the sample ID, type of measurement, and stage in the analysis process; these pieces of information will be separated by underscores. Examples: "MA006-20230901_TIMS_raw" and "CB012-20100512_SIMS_analyzed."

6. Do you need to do any version control on your files? How will that be done?

   Example: Version control will be very simple through file naming, appending analysis information onto the end of file names to keep track of which version of the file it is.

7. How will data move through the collection and analysis pipelines?

   Example: Once data is collected on the mass spectrometer, I will copy it to the correct folder on the shared server for analysis. Data will stay in its sample ID-labeled folder as it gets analyzed, with different file names to annotate analysis stage. Data that will be published will be copied into separate folders, organized by article.

8. Record any project roles and responsibilities around data management.

    Example: It is each researcher's responsibility to ensure that data moves through the analysis pipeline and is labeled correctly. The lab manager will ensure that the shared server stays organized and will periodically check that backups are working.

9. Write down any other details on how data will be managed.

    Example: Copies of this DMP will live in my top-level folder on the lab server so that others can find and use my data as needed.

10. Record all of this information in text file and save it in the top-level folder with your project files. Update the plan as necessary.

Exercise 5.3: Identify Applicable Data Policies

Data management and sharing are becoming expected and ingrained within the research process due to continually developing data policies from funding agencies, journals, and even individual institutions. While you might be aware of a policy from one of these sources, it's a good idea to check if any other policies require things from you and your data. By knowing about all applicable policies, you will be aware of any requirements around your data and can prepare for them, rather than being surprised if something goes wrong. However, because the policy landscape is so inconsistent and frequently changing, it's sometimes a challenge to know where and how to search for applicable policies. This exercise was designed to help with this searching and identification process.

This exercise prompts you to search specific places to look for relevant data policies and suggests terms to use during your search. You can do these searches in your favorite search engine, but you should also try the built-in search bar on your funder, journal, or institutional website, if available. Be aware that you might not find a policy in every step of this exercise because there is a chance it just doesn't exist. Once you've found a policy, write down the name of the policy and the important details so you know what is expected of you and your data.

1. Using your favorite search engine, search for:

    - [your funder name] and "data policy"
    - [your funder name] and "data management"
    - [your funder name] and "data sharing"

    Record any relevant policy titles and their key details:

    Example: NIH has the "Policy for Data Management and Sharing" which requires researchers to write a data management and sharing plan and to make data available in a data repository at the time of article publication or by the end of the grant period, whichever comes first.

2. Using your favorite search engine, search for:
   - [your institution name] and "data policy"
   - [your institution name] and "data stewardship"
   - [your institution name] and "data retention"

   If your institution has an internal portal where policies are stored, log in and search for:
   - "data policy"
   - "data stewardship"
   - "data retention"

   Ignore policies covering information technology (IT) or student data. Record any relevant policy titles and their key details:

   Example: Caltech does not have a research data policy.

3. If applicable, review your institution's Institutional Review Board (IRB)/ Ethical Review Board (ERB) policies and procedures for requirements around research data. Record details for any requirements:

   Example: This research has been deemed IRB exempt. IRB policies require controlled access to data and that researchers should also follow relevant funder policies.

4. If applicable, review your institution's technology transfer policies and procedures for requirements around research data. Record details for any requirements:

   Example: Not applicable.

5. If you have a Data Use Agreement (DUA) or Materials Transfer Agreement (MTA), describe any requirements for research data in those documents:

   Example: Not applicable.

6. Using your favorite search engine, search for:

- [your journal name] and "data policy"
- [your journal name] and "data availability"
- [your journal name] and "data sharing"

Record any relevant policy titles and their key details:

Example: PLoS has a Data Availability Policy which requires that authors make data available, preferably in a data repository, to replicate the results of any research published in their journals.

7. If you know of any other policies that affect the management and sharing of your research data, identify those policies and record their key details:

Example: Not applicable.

Exercise 5.4: Determine Data Stewardship

Expectations for data use during a project are often clear but become murky after a project ends. For example, if the project ends and you leave the research group/institution, do you have permission to reuse the project data? Who is responsible for keeping the data after the project ends? What about keeping physical research notebooks? Not knowing the answer to these questions can lead to confusion, hurt feelings, loss of data, or even article retraction (McCarty 2021a, 2021b). This is why it is recommended to be up front about requirements and permissions around research data before a project ends.

This exercise encourages you to discuss data permissions and stewardship with supervisors and peers to make sure that there are no misunderstandings about who has what rights to use, retain, and share data. To complete the exercise, first determine which research data should be discussed. Bring together the Principal Investigator, the researcher collecting the data, and anyone else who works with that data. As a group, answer the questions in the exercise, making sure that everyone agrees on the final decisions. Record the results of the discussion and save them with the project files.

This exercise was adapted from the "Project Close Out Checklist" (Briney 2020c).

1. Who is participating in this discussion?

    Example: This discussion includes the graduate student who collected the data, the project Principal Investigator (PI), and the laboratory manager.

2. What data is being discussed?

   Example: This discussion covers all of the data collected by the graduate student during their time at the university.

3. Are there security or privacy restrictions on the data and, if so, what are they?

   Example: Some of the research data includes human subjects data. This data must be held securely with limited data sharing, as outlined in the IRB protocols.

4. Are there intellectual property limitations on the data and, if so, what are they?

   Example: There are no intellectual property concerns for the data.

5. Are there any requirements to publicly share the data and, if so, what are they?

    Example: This research was funded by the NIH, which requires data sharing. The laboratory plans to share all data reproducing published results with the exception of the human subjects data.

6. If data will be publicly shared, who will do so and what are the parameters for sharing?

    Example: The lead author on each manuscript will share any data supporting the article that can be made publicly available. Credit via data authorship will be given to all article co-authors plus anyone who assisted with data collection and analysis.

7. Who will store the copy of the record of the data and for how long?

    Example: The project PI will retain the copy of the record of the data for at least 3 years after the end of the grant award, with an ideal 10-year retention period.

8. Who is allowed to keep a copy of the data after the project ends? Which data?

   Example: The graduate student may keep a copy of all data except the human subjects data after they leave the university.

9. Who is allowed to reuse the data after the project ends? Which data? Are there any requirements for reuse, such as co-authorship?

   Example: The graduate student may reuse and publish with the data collected during their time at the university but must offer co-authorship of any papers using the data to the project PI and any relevant lab members.

10. Who keeps any physical research notebooks after the project ends?

    Example: The PI will keep all physical laboratory notebooks but the graduate student may make copies to retain for their personal records.

11. Make a copy of this information and save it alongside the relevant data files.

6

DATA SHARING

As part of my job as a librarian, I regularly teach about data sharing. In the early 2010s, this involved a lot of basic messaging that researchers were now expected to share their data (with "sharing by request" still an allowable mechanism at that point). My guidance for data sharing has evolved significantly over the years. I now teach that data should be shared in a data repository, be shared with a DOI or other permanent identifier, have clear reuse permissions or restrictions (e.g. for sensitive data), and be documented to the point where someone with similar training could understand the data. The landscape for sharing research data has changed so much in two decades – and is still changing – that it's not surprising that researchers often struggle with following the most current recommended or required practices for sharing. Not to mention that, as I know from making my own research data publicly available, it can take significant time to prepare your data for sharing. The good news is that there are resources, such as the exercises in this chapter, to help guide you through the process until data sharing practices become comfortably ingrained in your research workflows.

Sharing data that underlies research is now a common expectation within scholarly research, but the landscape of data repositories is still uneven and funding agencies come out with new data sharing policies every few years. This is why you will find guidance on the basics of data sharing, picking a data repository, and writing good documentation in this chapter. We are still in a time of changing expectations, so it is good to be comfortable with the fundamentals. Beyond this, the chapter contains a pair of exercises to make sure your data and scientific figures are maximally understandable and reusable. There is not as much emphasis currently on accessibility and reusability over simply making data available, but this is an important part of equitable access to research. Information should be made available for everyone, including disabled people, and not limited to researchers who can afford specialized software or the able-bodied. There's so much potential for making research available and accessible to everyone, so I encourage you to think broadly about accessibility and reusability when sharing research data.

Exercise 6.1: Make a Spreadsheet more Accessible and Reusable

Many data management actions are prompted by the desire for greater transparency and reusability. For example, making your data both accessible and reusable makes it easier for someone – including your future self – to use and understand your data. Big differences in reusability can even come from slight tweaks to formatting a dataset. As spreadsheets are one of the most common data types, this checklist provides guidance on making a spreadsheet reusable as well as more accessible to those with disabilities. Note that sometimes guidance for reusability and accessibility may be in conflict, and so this checklist is a best effort to balance the two considerations.

For a given spreadsheet, complete the exercise by working through the actions on this checklist to make the data more accessible and reusable. This is usually done when the data is finalized and/or prior to sharing the data either publicly or with colleagues. If you must share the data in its original form, be sure to make the accessible version available alongside it.

For more guidance on spreadsheet accessibility and reusability, see Wickham (2014), Broman and Woo (2018), and Oxford and Woodbrook (2023).

Organize the data:

- ☐ Break data into several smaller rectangular tables instead of one large complex table, as necessary. Each sheet should contain only one table.
- ☐ Arrange data so that the first row of the spreadsheet contains variable names, with data of all following rows starting in the first column.
- ☐ Start the table in cell A1; there should be no empty columns or rows in the top left of the spreadsheet before the table starts.

Make the data readable:

- ☐ Clean up the variable names in the first row of the spreadsheet to be both human and machine readable:
  - ☐ Use short but meaningful and distinct variable names.
  - ☐ Use full words or readable abbreviations (e.g. "number" or "num" instead of "n") in variable names.
  - ☐ Use only alphanumeric characters in variable names.
  - ☐ Remove spaces from variable names.
  - ☐ Format all variable names in either "camelCase", "PascalCase", or "pothole_case":
    - ☐ camelCase: capitalize the first letter of each word in the variable name except for the first one (e.g. myVariableName).

□ PascalCase: capitalize the first letter of each word in the variable name, including the first (e.g. MyVariableName).

□ pothole_case: separate each word with an underscore (e.g. my_variable_name).

□ Place the key, or most identifying, variable in the first column on the left, column A. (Spreadsheets should be readable from left to right then top to bottom, and placement of the key variable in the first column will help with readability.)

Clean the data:

□ Convert any dates to YYYY-MM-DD format. (To work around Excel's weird date formatting you can separate year, month, and day into three separate variables.)

□ Ensure that spreadsheet cells contain only one data point. If there is more than one data point per cell, divide columns into multiple variables, as appropriate.

□ Remove formatting such as font, text alignment, highlighting, and merged cells. Any information that is represented by such formatting should be recorded as data under new variable.

□ Remove color from the spreadsheet, leaving text black on a white background for maximum contrast.

□ Fill in empty cells:

□ input any missing data values;

□ use "NA" (or the preferred null value for your analysis software) for any cells that do not have recorded values.

□ Perform quality control on the data, removing:

□ errors;

□ inconsistencies;

□ accidental spaces.

Format the data:

□ Remove charts. Charts may be shared separately with corresponding alt text (see Exercise 6.3: Write alt text for a data visualization).

□ Remove underlying calculations so that the file only includes raw data. (You can do this in Excel by copying a column, using the special paste option to "paste as values", then deleting the original column.)

□ Use any built-in validation or accessibility checkers provided by your software.

Save and share the data:

- ☐ Save data as a CSV file type (TSV is also an acceptable file format). Save individual spreadsheet tabs as separate CSV files.
- ☐ Give your file a descriptive file name (see Exercise 3.2: Create a file naming convention).
- ☐ Create an accompanying data dictionary (see Exercise 2.3: Create a data dictionary).
- ☐ Share the accessible CSV file(s), the data dictionary, and, as needed, the original spreadsheet.

Exercise 6.2: Write a README File for Shared Data

A README file can make the difference between not understanding a shared dataset and actually being able to reuse the data. A README file for shared data goes beyond basic information listed in a data repository by explaining what is in each file, how files relate to each other, and anything else someone needs to know to use a dataset. While you may not be required to include a README file when depositing data into a data repository, including a README.txt file (or README.md for shared software) is almost always recommended. (Note: repositories for very specific types of data (e.g. genetic data) sometimes do not accept README.txt files and instead ask for similar, detailed information during the deposit process.)

This exercise walks you through creating documentation for data that is being shared, whether that is one file or many. To complete the exercise, answer the questions for the collection of data being shared. Save the information to a TXT file with the name "README" and share it with your data. (The name, "README," indicates that the file conveys important information and the file type, TXT, can be opened by many different software programs, making the content maximally accessible.) Each group of files being deposited into a repository should have an accompanying README file, meaning you should do this exercise more than once if you are depositing data in multiple groups/repositories.

This exercise was developed for *The Data Management Workbook* and includes several items from "Best Practices for Data Submission in Generalist Repositories: A Checklist" (Curtin et al. 2025).

1. Record the title of the dataset:

   Example: Data from "Measuring data rot: an analysis of the continued availability of shared data from a single university"

2. List the dataset authors, identifying the corresponding author(s) and providing their contact information.

   Example: Kristin Briney (ORCID: 0000-0003-1802-0184)

3. Briefly summarize the project these data files are from.

   Example: This data is from a study of 2,000+ links to shared data from Caltech-authored publications. The links were web-scraped to test for the continued availability of the data.

4. If applicable, list any publications supported by this data.

   Example: One article was published from this data: Briney, K. A. (2024). Measuring data rot: An analysis of the continued availability of shared data from a Single University. PLoS ONE, 19(6), e0304781. https://doi.org/10.1371/journal.pone.0304781

5. If applicable, record the funding source for this data.

    Example: The author received no funding for this work.

6. Identify the license or permissions this data is being made available under (see Exercise 6.5: Select a license for your shared data).

    Example: The data is available under a Creative Commons Zero v1.0 Universal license.

7. Write a list of all data files being shared and a short description of what each of these files contains.

   Example:

   - *DataRot.csv: This data contains all of the links tested, listing results of the web-scraping but not results of the hand testing.*
   - *DataRot_dataDictionary.txt: Data dictionary defining variable names and values for DataRot.csv.*
   - *DataRot_handTested.csv: Subset of supplemental data links from DataRot. csv that were hand-tested and the results of the hand testing.*

8. Note any specialized software or tools needed to access or use the data.

   Example: No specialized software is needed to access the data files as they are all in open formats. The data is UTF-8 encoded. R version 4.1.1 was used for web-scraping and data analysis.

9. Describe any relationships between files.

   Example: DataRot_handTested.csv is a subset of DataRot.csv that only includes the links that were tested by hand.

10. Is there any related data or content not shared here? If so, list that content, describe it briefly, and document how it can be accessed.

    Example: The code used for web-scraping and analysis of this data is available at https://doi.org/10.22002/d2h9g-5q152 under a GNU General Public License v3.0.

11. For any spreadsheet files, create a data dictionary (see Exercise 2.3: Create a data dictionary). Either copy the data dictionary contents here or share the data dictionary as a separate file (making sure to add the data dictionary to the above list of files and their descriptions).

    Example: The data dictionary is available in the DataRot_dataDictionary.txt file.

12. What else should someone know about this dataset?

Example: This is version 2 of the data, which reflects changes based on reviewer feedback during peer review of the related article.

13. Save all of this information as a README.txt file (alternatively save it as README.md when sharing software), format it for clarity, and share with your data.

Exercise 6.3: Write Alt Text for a Data Visualization

Writing alt text (short for "alternative text" and means a text description of an image) is one of the most basic ways to make a scientific figure accessible to a broader audience. Alt text is a basic requirement when sharing digital images so they are understandable to blind people using screen readers, to help with search engine optimization, and to provide textual information whenever an image does not load on a webpage (such as if you are doing fieldwork and have a poor internet connection). Alt text is useful when sharing figures on your research group website, on social media, and even when publishing journal articles.

This exercise will help you create basic alt text for a data visualization. To complete the exercise, answer the prompts for a selected figure. An example is provided based on Figure 6.1. For figures with multiple panels, answer

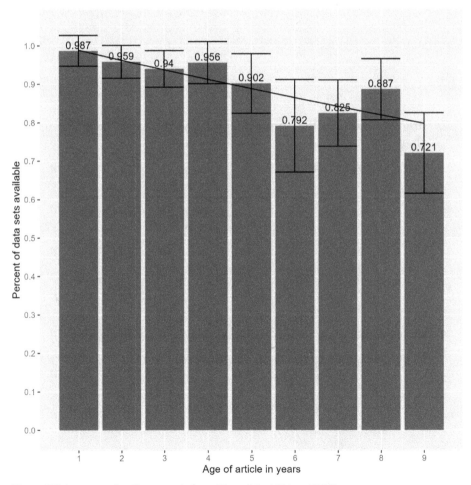

Figure 6.1: Image used as the example for writing alt text (Briney 2024).

questions 1–4 for each panel of a figure then assemble the answers in step 5. This formula for creating alt text comes from Amy Cesal (Cesal 2021) writing for the Data Visualization Society. It's less accessible than a more complete description of a visualization, but it's a quick way to write usable alt text for a chart. Once you've written the alt text, be sure to share it any time you share the figure; the method for adding alt text to an image varies by platform.

For more complete guidance on writing alt text, see DIAGRAM Center (2019), Schwabish, Popkin and Feng (2022) and GBH (2024).

1. In two or three words, list the chart type:

    Example: Column chart

2. Briefly summarize the primary data (e.g. y-axis data) that makes up this chart:

    Example: Research data availability

3. In one sentence, write the main takeaway of the visualization:

    Example: Research data on the internet disappears at a rate of 2.6% per year.

4. Write out the full alt text as: [answer 1] of [answer 2] where [answer 3].

   Example: Column chart of research data availability where research data on the internet disappears at a rate of 2.6% per year.

5. Where applicable, repeat steps 1–4 for each panel of a multi-panel figure. Note which part of the figure goes with each sentence.

   Example: Not applicable

6. Where possible, include a link to the data underlying the figure.

   Example: For underlying data, see "Figure2_UnavailableByYear.csv" file at https://doi.org/10.22002/h5e81-spf62

7. Write down the final alt text, combining the answers from steps 4–6.

   Example: Column chart of research data availability where research data on the internet disappears at a rate of 2.6% per year. For underlying data, see "Figure2_UnavailableByYear.csv" file at https://doi.org/10.22002/h5e81-spf62

8. Share the alt text anytime you share the figure.

Exercise 6.4: Pick a Data Repository for Sharing

It is sometimes difficult to find the best data repository for sharing data, as many data repositories exist but coverage is uneven across disciplines. While there is a complete list of available data repositories (re3data 2025), it can be overwhelming to navigate. When I help a researcher find the best repository for a dataset, I use a different strategy: start by looking for known repositories for specific types of data, then seek out broad disciplinary data repositories, and, if a suitable data repository has not yet been found, default to a generalist data repository. There are a couple other considerations in this search. The first is that repositories sometimes charge fees for deposit; a few repositories charge for all deposits and most repositories charge for large data (500 GB or larger), if they accept large data at all. Be sure to budget appropriately. The other consideration is whether the repository will give you a DOI or similar permanent identifier, as is required under the most recent data sharing guidance and funder mandates. If the repository does not provide you with an identifier, such as a DOI or accession number, consider depositing your data in a different repository if you can.

This exercise walks you through a decision tree (visually represented in Figure 6.2) for selecting an appropriate data repository. An example is provided for data and code from an article I wrote (Briney 2024). To complete this exercise, identify the data that needs to be shared and work through repository selection from discipline-specific data repositories to more general data repositories. Once you have identified a repository for all of your data, deposit the data and skip to step 8 to record the corresponding permanent identifiers. Note that, depending on data types, you may need to deposit your data into multiple repositories (for example, a discipline-specific repository for one type of data and an institutional data repository for the rest of the data).

1. **Identify all of the data that needs to be shared.**

   Example: The content to be shared includes code and data from analysis of the continued availability of shared research data on the internet.

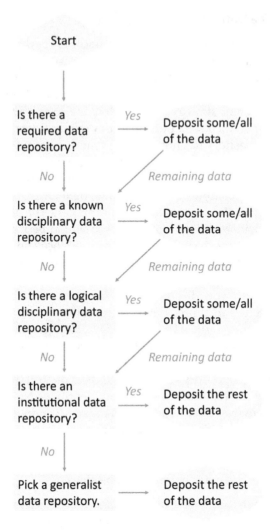

Figure 6.2: Workflow diagram upon which the exercise is based. Starting at the top left: identify if there is a required repository (if so, deposit data), decide if there is a known disciplinary data repository (if so, deposit data), a logical disciplinary data repository (if so, deposit data), or an institutional data repository (if so, deposit data), and if none of those work pick a generalist data repository.

2. Is there a required data repository for some or all of the data? For example, is there a data repository required by your funding agency?

   If so, deposit some or all of your data there. Go to step 8 if the repository will accept all of your data or go to the next question if there is still some data left to deposit.

   Example: There is no required repository for this data.

3. Is there a known disciplinary data repository for some or all of the data? For instance, is there a data repository used by everyone in your research area?

   If so, deposit some or all of your data there. Go to step 8 if the repository will accept all of your data or go to the next question if there is still some data left to deposit.

   Example: Scientific code is usually shared on GitHub for easier reuse, so the code will be shared there. However, GitHub does not provide permanent identifiers.

4. Find a list of recommended data repositories from a journal, such as those from *PLoS* (PLoS ONE 2023) or *Springer Nature* (Springer Nature 2025). Is there a logical disciplinary data repository for some or all of your data?

   If so, deposit some or all of your data there. Go to step 8 if you have shared all of your data or go to the next question if there is still some data left to deposit.

   Example: There isn't a logical disciplinary data repository for this data.

5. Does your institution have a data repository?

   If so, deposit the remainder of your data there and go to step 8.

   Example: The California Institute of Technology hosts the data repository CaltechDATA. I will deposit my data in CaltechDATA and also put a copy of my code there, in order to get DOI's for both.

6. Do you have a preferred generalist data repository (e.g. Dryad, Dataverse, figshare, Open Science Framework, Zenodo, etc.)?

   If so, deposit the remainder of your data there and go to step 8.

   Example: (All data has been shared already.)

7. Pick a generalist data repository using the Generalist Repository Selection Flowchart (Barbosa et al. 2024) and deposit the remainder of your data.

   Deposit your data and go to step 8.

   Example: (All data has been shared already.)

8. Record the permanent identifier, ideally a DOI, from each data deposit.

   DOIs make data FAIR (Wilkinson et al. 2016) and aid with data sharing compliance. If you did not receive a permanent identifier (such as a DOI, permanent URL, accession number, etc.) during deposit, return to step 2 and pick a different data repository for your data.

   Example: The data is in CaltechDATA at https://doi.org/10.22002/h5e81-spf62 and the code at https://doi.org/10.22002/d2h9g-5q152. The code is also available on GitHub at https://github.com/kbriney/SupplementaryData, which is the community-accepted location for sharing scientific code even though there is no DOI and the content is not unalterable.

Exercise 6.5: Select a License for Your Shared Data

Data sharing can be overwhelming if you have never done it before, particularly when it comes time to select a license for your shared data. If you are not an expert at data and copyright (and very few people are), the different license options can be confusing. Plus, it's not always clear at a glance when to use one license over another. To provide some clarity in this area, this exercise walks you through a flow chart to determine the best licensing option for your data.

Be aware that the default license in this flowchart is CC0, the public domain waiver. This is the recommendation of multiple groups (Murray-Rust, Peter et al. 2010; Creative Commons Wiki 2014) because most research data are not actually copyrightable, which means using a license that leverages copyright is not appropriate. Additionally, CC0 maximizes reuse by avoiding attribution stacking (combining multiple datasets with licenses that don't mesh together); attribution stacking can make it impossible to legally reuse and give credit to all of the chosen data. There are also times when another license or data sharing mechanism is more appropriate, such as using an Open Data Commons licenses for a shared database or a Data Use Agreement for sensitive data, which are accounted for in the exercise. Finally, no matter what license is chosen, there is a moral obligation for the data re-user to cite the original data, even if attribution is not explicitly required by the license; see Exercise 6.7: Cite a publicly available dataset to create a citation for a dataset.

To complete this exercise, work through the decision tree (visually represented in Figure 6.3) until you have selected an appropriate license/sharing mechanism, then jump to step 7 to record the chosen license/sharing mechanism. Be aware that this exercise is only for choosing a license for shared datasets, and should not be used for choosing a license for articles, shared software, or any other content type outside of data.

1. Identify the data that will be shared:

    Example: The data is a spreadsheet consisting of observations of bald eagles in the San Bernardino National Forest in California.

Start

Are there ethical limitations on data reuse? — *Yes* → DUA/controlled-access repository

No

Is the data formatted as a database? — *Yes* → Use an Open Data Commons license

No

Does the data repository require a license? — *Yes* → Use the required license

No

Use the CC0 waiver

Figure 6.3: Workflow diagram upon which the exercise is based. Starting at the top left: determine if there are ethical limitations on data reuse (if so, use a Data Use Agreement and/or a controlled-access repository); ascertain if the data is formatted as a database (if so use an Open Data Commons license); identify if the chosen data repository has a default license (if so, use that license); and if no license has previously been selected, choose the CC0 waiver.

2. Do you have the rights to share this data?

   If you did not create the data, examine the license or Data Use Agreement (DUA) the data was received under to learn if you have permission to share further. If you created the data, see Exercise 5.4: Determine data stewardship to ascertain permissions. If you have the rights, continue the exercise. If you do not have the rights, end the exercise and do not share the data.

   Example: The data was collected within my research lab and I have permission to share so long as my PI and coworkers are credited as data authors.

3. Do there need to be limitations on data reuse for legal, ethical, or other reasons and what would they be?

   If there are reuse limitations but the data still needs to be shared, consider using a Data Use Agreement (DUA) instead of an open license and/or a data repository with controlled access. If either of these options are chosen, advance to step 7.

   Example: This data does not contain any human subjects information or patent information. Location data has been obscured in this dataset to protect the observed animals, which means that there are no limitations on reuse.

4. Is the data formatted as a database?

   If yes, select an Open Data Commons (ODC) license (Open Knowledge Foundation 2025), as they specifically account for database rights. Use either the Open Data Commons Attribution License or the Open Data Commons Public Domain Dedication and License (PDDL). Advance to step 7.

   Example: The data is not a database.

5. Is there a default/required license for the data, such as from the data repository?

   If yes, use that license. Advance to step 7.

   Example: The data will go into the Dryad repository, which requires a CC0 waiver for all deposited data.

6. If you have not already selected a license for data sharing, default to the CC0 waiver for your data.

   The CC0 waiver removes any reuse limitation on shared data by placing it in the public domain, though there is still the moral obligation to cite the data if it is reused. Continue to the next step.

   Example: (A license has already been selected.)

7. Write down the chosen license or terms of sharing for your data.

   Example: I will use the CC0 waiver for my data.

Exercise 6.6: Share Data

Data sharing has become common and expected by funding agencies and journals, but it is still a relatively new practice within many disciplines. For this reason, it is worth being sure you're always carrying out the necessary steps when making data publicly available. The process of depositing data into a data repository will vary between repositories, but there are some common actions that should be taken to prepare data for sharing. This exercise walks you through these standard requirements for sharing data.

This exercise is a checklist that details the necessary steps and decisions to make when depositing data into a data repository. To complete the exercise, identify the data to be shared and work through the actions on the list. Note that, if data will be shared as multiple deposits or in multiple repositories, the checklist should be worked separately for each group of deposited data. If you are depositing a large number of datasets, it can also be worthwhile to ask repository administrators about potentially using an Application Programming Interface (API) to skip manual entry of duplicate metadata.

This exercise was developed for *The Data Management Workbook* and includes a few items from "Best Practices for Data Submission in Generalist Repositories: A Checklist" (Curtin et al. 2025).

Select data:

- ☐ Select the data files that, at minimum, reproduce published results.
- ☐ Identify any other content that needs to be shared for reproducibility, including code, survey instruments, etc.
- ☐ Prepare any sensitive data for sharing, such as by de-identifying the data, obscuring sensitive data points, etc. Consider if sensitive data should be shared under a Data Use Agreement (DUA) and/or via a controlled-access data repository instead of through public sharing.
- ☐ Perform quality control on the data files (see Exercise 6.1: Make a spreadsheet more accessible and reusable and refer to disciplinary norms for data cleaning).
- ☐ Convert data from proprietary file types to open file types, as appropriate (see Exercise 7.1: Convert data file types).
- ☐ Determine if data should be shared under one citation or as several citations. This may be governed by data repository restrictions.
  - ☐ If there are no repository restrictions, group data as makes most sense for citation and reuse. Options can include: sharing as one large group, grouping files by data type, giving large data files their own citations, etc. Each citation represents a unique deposit into a data repository.

☐ If there will be multiple deposits in one repository or data will be divided across more than one data repository, work through the remainder of the checklist separately for each citation/group of files.

Define sharing information (metadata):

☐ Give the dataset a title. Default title is "Data from: [name of the article]."

☐ Write a brief description of the dataset to be used as the abstract/description.

☐ Write down keywords/subject terms for the data. Default to the keywords from the associated publication.

☐ Determine who will be listed as authors of the data and in what order; this may be different than the authors of the article.

☐ Identify author ORCID numbers for submission (note: this is best practice, but not all data repositories have integrated ORCID (ORCID 2025), which is a system for uniquely identifying authors and facilitating proper attribution).

☐ Record all funding information that contributed to the dataset.

☐ Chose a license for reuse rights (see Exercise 6.5: Select a license for your shared data). The default recommended license is CC0.

☐ Note the DOIs/URLs of any related content – such as code, additional data, preprints, etc. – that is associated with the data.

Document data:

☐ Write a README.txt file to be shared with the other files (see Exercise 6.2: Write a README for shared data).

☐ Document any spreadsheet data with a data dictionary (see Exercise 2.3: Create a data dictionary). The data dictionary should be shared with the other files.

Deposit data:

☐ Pick a data repository or data repositories for the shared data (see Exercise 6.4: Pick a data repository).

☐ Deposit the data and documentation files into the data repository, and fill in metadata as determined in the "Define sharing information" section.

Share data:

☐ Record the data's DOI or other permanent identifier, such as accession number.

☐ Link the publication to its data, usually through a Data Availability Statement.

Exercise 6.7: Cite a Publicly Available Dataset

When you use a publicly available dataset, the proper way to give credit is to cite it as you would a journal article: briefly in-text, then with the full citation information in a bibliography. This method of citation adds to the dataset's citation count, a metric that can be used to show the impact of that data. You can also add information on the dataset to the "Data Availability Statement" section of an article but this is for transparency and does not contribute to citation counts for the reused data. With citation as the preferred pathway for giving credit for data reuse, we now need to worry about proper citation formatting for this content type.

There has been significant work done to develop standards for citing datasets within various style guides. For example, the ACS (Vogel et al. 2025), APA (APA 2020), and ICMJE (Patrias and Wendling 2018) style guides have specified formats for dataset citations, and Chicago and MLA styles have community-accepted standards (IASSIST SIGDC 2012; Purdue University 2024). Additionally, repositories sometimes offer automatically generated citations for the datasets they contain, making it easy to copy-paste a citation into a document (see the citation from Dryad in Table 6.1, which is the repository that hosts this exercise's example dataset). While not universal to all style guides, there are now many options for formatting dataset citations.

This exercise guides you through creating and formatting a dataset citation for several citation styles with identified formats for citing data. The exercise provides a formatting template and an example citation for a dataset by Zehr et al. (2019) in each of five citation styles, plus the auto-generated citation from the original repository. The exercise also works through formatting a citation for a different dataset from Vines et al. (2014). If your chosen citation style is not shown in this exercise, try formatting your citation using a citation manager, though be aware that citation managers don't always handle dataset citations accurately (Vrouwenvelder et al. 2025). Once your citation is formatted, be sure to cite the dataset in your document as you would for an article or book.

1. Gather information about the dataset.

   Record the following list of information about the dataset to incorporate into the data citation. Note: Permanent identifiers come in many types; in preferential order, you should use the first available of the following: 1) the dataset DOI; 2) an ARK, Handle, or PURL for the dataset; 3) the URL that links directly to the dataset's information page; and 4) the URL for the repository/website and an accession/identifier number for the data, if one is available.

Information	Example Dataset	Your Dataset
Dataset title	*Data from: The availability of research data declines rapidly with article age*	
Author/creator name(s)	*Timothy H. Vines, Arianne Y. K. Albert, Rose L. Andrew, Florence Débarre, Dan G. Bock, Michelle T. Franklin, Kimberly J. Gilbert, Jean-Sébastien Moore, Sébastien Renaut, & Diana J. Rennison*	
Date published	*2014-11-05*	
Version number, if applicable	*[not applicable]*	
Publisher (repository/ website name)	*Dryad*	

Information	Example Dataset	Your Dataset
DOI/permanent identifier for the dataset	*doi: 10.5061/dryad.q3g37*	
Data accessed	*2025-03-17*	

2. Determine your reuse rights to the data.

If the data has an identified license, follow the terms of that license. If the identified license is too restrictive, contact the dataset creator to ask about broader reuse rights. You may have other reuse rights depending on copyright laws in your country; when in doubt, contact a copyright expert for guidance.

Example: The data is available under the CC0 1.0 Universal public domain waiver, meaning there are no reuse restrictions.

3. Determine your citation style.

This exercise provides formatting guidelines for data citations in ACS, APA, Chicago, ICMJE, and MLA citation styles.

Example: I will use APA citation style.

4. Format the citation according to your chosen citation style.

Take the recorded information about the dataset from step 1 and place it in the proper arrangement for your chosen citation style. Follow the formatting templates outlined in Table 6.1. If none of the templates in Table 6.1 match your citation style, try formatting the data citation with a citation manager.

Table 6.1: Data citation templates and examples for five citation styles and an auto-generated citation from the Dryad repository.

Citation Style	Template	Example
ACS	Author(s). *Dataset Title* (Dataset), Version. Publisher, Date published. DOI/URL	Zehr, S. M.; Roach, R. G.; Haring, D.; Taylor, J.; Cameron, F. H.; Yoder, A. D. *Data from: Life History Profiles for 27 Strepsirrhine Primate Taxa Generated using Captive Data from the Duke Lemur Center* (Dataset), Ver 2. Dryad, February 18, 2019. DOI: 10.5061/dryad.fj974
APA (7th edition)	Author(s). (Date published). *Dataset title* (Version) [Data set]. Publisher. DOI/URL	Zehr, S. M., Roach, R. G., Haring, D., Taylor, J., Cameron, F. H., & Yoder, A. D. (2019). *Data from: Life history profiles for 27 strepsirrhine primate taxa generated using captive data from the Duke Lemur Center* (Version 2) [Data set]. Dryad. https://doi.org/10.5061/dryad.fj974
Chicago (17th edition) author-date	Author(s). Date published. *Dataset Title*. Version. Distributed by Publisher. DOI/URL.	Zehr, Sarah M., Richard G. Roach, David Haring, Julie Taylor, Freda H. Cameron, and Anne D. Yoder. 2019. *Data from: Life History Profiles for 27 Strepsirrhine Primate Taxa Generated using Captive Data from the Duke Lemur Center*. V. 2. Distributed by Dryad. https://doi.org/10.5061/dryad.fj974.

Citation Style	Template	Example
ICMJE	Author(s). Dataset title. Date published [amended version date; cited citation date]. In: Publisher [Internet]. Place of publication: Publisher's organization. Publisher active date(s). Dataset size. Available from: DOI/URL	Zehr SM, Roach RG, Haring D, Taylor J, Cameron FH, Yoder AD. Data from: Life history profiles for 27 strepsirrhine primate taxa generated using captive data from the Duke Lemur Center. 2015 Jun 26 [amended 2019 Feb 18; cited 2025 Mar 4]. In: Dryad [Internet]. Durham (NC): Dryad. 2008 - . 13.07 MB. Available from: https://doi.org/10.5061/dryad.fj974
MLA (9th edition)	Author(s). *Dataset Title*. Version, Publisher, Date published, DOI/URL.	Zehr, Sarah M., et al. *Data from: Life History Profiles for 27 Strepsirrhine Primate Taxa Generated using Captive Data from the Duke Lemur Center.* Version 2, Dryad, 2019, doi:10.5061/dryad.fj974.
Dryad repository (auto-generated)		Zehr, Sarah M.; Roach, Richard G.; Haring, David et al. (2019). Data from: Life history profiles for 27 strepsirrhine primate taxa generated using captive data from the Duke Lemur Center [Dataset]. Dryad. https://doi.org/10.5061/dryad.fj974

Write your formatted citation here:

Example: Vines, T. H., Albert, A. Y. K., Andrew, R. L., Débarre, F., Bock, D. G., Franklin, M. T., Gilbert, K. J., Moore, J-S., Renaut, S., & Rennison, D. J. (2014). *Data from: The availability of research data declines rapidly with article age* [Data and code]. Dryad. https://doi.org/10.5061/dryad.q3g37

5. Add the data citation to your article bibliography and cite the dataset in-text as you would for an article.

7

PROJECT WRAP-UP

I used to hate it when my boss frantically emailed me asking for data or a figure from an older, completed project, telling me they needed the file in the next two hours. This has happened to me several times and it inevitably meant performing an archeological excavation inside computer folders I hadn't touched in years. More recently, I've started doing specific data management tasks around project wrap-up that, on top of strong file organization and naming systems, has made finding older data and files infinitely easier. In particular, setting aside key files at the end of a project in an "archive" folder has made it so that I can usually find the necessary file within five minutes. I sometimes still have to do deep dives into older data and documentation, but I know without a doubt that I can always find the correct version of a dataset used to make the figure – plus the figure itself – in my archive folder. So while I still dislike last-minute requests for data from my boss, it has become much easier to deal with them through strategic data management.

The end of a project is a key time to perform data management activities in order to set yourself up for future data reuse. This is because you still remember all of the important details about your data and can make good decisions about preparing it for the future. A lot of this is just making sure your data is in order and well documented, but specific tasks such as converting file formats and setting aside the most important files into an "archive" folder, can make the difference between easy future reuse and a data nightmare. Beyond a project wrap-up checklist and guidance for file formats and an archive folder, this chapter also addresses a very common project wrap-up task: leaving your institution. As I work in academia, this situation arises frequently as students and post docs, by their very nature, stay at universities for a finite amount of time. Hopefully, the exercises in this chapter will help ease the stress of such frequent transitions and make it easier to pick up your research anytime you need to revisit older projects.

Exercise 7.1: Convert Data File Types

Data is often stored in a file type that can only be opened by specific, costly software – this is referred to as a "proprietary file type." An easy way to tell that you have data in a proprietary file type is if you cannot open the data file when you lose access to the original software. When data is in a proprietary file type, it's always a good idea to copy that data into a more common, open file type that many software programs can open. This copying can result in some loss of functionality like formatting, but it's better to have a backup that you can use than to not having any data at all if you don't have access to the proprietary software anymore.

This exercise works through identifying possible alternative file types for the data's proprietary file type before instructing you to make a copy of the data in the new file type. To do the exercise, identify the data to convert and answer the questions. Once you have picked a more open, common file type, make a copy of the data in that file type but do not delete the original data. (Keeping a copy in the original file format means that, so long as you have access to the necessary software, your data has full functionality. If you lose access to the software, you'll still have your data in some format, which is better than not having your data at all.) The exercise should be done before sharing any data in a proprietary file type.

1. What data is being converted?

   Example: The data is microscope images of zebrafish in the .CZI file format.

2. Is your data stored in a proprietary file type? What file type and how does this limit future data reuse?

   Example: Data is stored in a .CZI file format, which is a proprietary Zeiss microscope image format. These files do not open in other software.

3. Is it possible to convert your data to other file types? If so, list the possible types:

   Example: I can use the Bio-Formats tool to convert .CZI files to: .AVI, .CH5, .DCM/.DICOM, .EPS/.EPSI/.PS, .ICS/.IDS, .JPG, .JP2/.J2K/.JPF, .MOV, .OME.TIFF/.OME.TIF, .OME/.OME.XML, .PNG, or .TIFF/.TIF.

4. Which of the possible file types are in common use? Which of the possible file types can be opened by multiple software programs?

   Example: JPG, PNG, and TIFF are all image formats in common use. OME-TIFF is a common image format within microscopy; most software will read the TIFF portion of the file but only some software will read the extra OME metadata. Common movie file types are AVI and MOV.

5. Of the possible options above, do you have a preference for specific file type(s)?

    Example: I prefer an image file over a movie file. TIFF is best because it doesn't lose resolution due to compression and can store all of the 4-dimensional image layers. OME-TIFF gives all of the benefits of TIFF but with added metadata.

6. Pick one of the more open or common file types and copy your important data files into that file type. Do not delete the original files.

    Example: I will convert my data to OME-TIFF files.

Exercise 7.2: Create an "Archive" Folder

One of the most useful data management tasks is, at the end of a project, to set aside key research and data files in an "archive" folder. The archive folder should only contain a small subset of your documents, usually the ones most likely to be reused. This will help you save time later when you need to find important files from the project quickly. It's best to create this archive when you end the project while you still remember which files are important and where they are located. You may still need to go through all of your files later to find a specific piece of information but, in the majority of instances, you will save time because the file you're most likely to need is in the archive folder.

This exercise consists of a checklist of the key documents that are likely to be most useful in a research project archive. Create a separate folder within the larger project folder (or in a highly visible place within the storage system) labelled "archive." Copy – do not move – the files on this checklist into the archive folder. Add copies of other important research documents, as needed. Remember, the archive folder does not need to be comprehensive, so focus on the subset of files that are most likely to be reused or referenced in the future.

This exercise was adapted from the "Project Close-Out Checklist" (Briney 2020c).

Project documentation:

☐ The project information file (see Exercise 2.2: Document project information).

Data snapshots:

☐ Important raw data.

☐ Key data analyses.

☐ Final published data.

Code:

☐ Analysis code.

☐ Record software versions, as appropriate.

Other research documents:

☐ Protocols (e.g. experimental protocols, IRB/ERB and animal care protocols).

☐ Survey instruments.

Research notes:

☐ Scans of research notebook.

☐ Digital notes.

Images:

- ☐ Flat files of figures (e.g. .JPG or .TIFF).
- ☐ Editable image files (e.g. .XLSX or .PSD).

Publications:

- ☐ Published articles in .PDF format.
- ☐ Accepted version of articles in editable document format (e.g. .DOCX).
- ☐ Poster files.

Administrative documents:

- ☐ Grant proposals.
- ☐ Grant progress reports and final report.

Exercise 7.3: Prepare Data for Future Use

The end of a project is a good time to prepare data for potential future reuse, as you still know the important details about the data to record and have access to any software used to create the data. End of project activities include tasks like converting file formats and making an archive folder (as seen in the previous two exercises) as well as organizing and documenting your data. The goal is to have well documented project data in one central location that is organized and formatted in a way to make future reuse easier.

This checklist exercise walks you through steps to gather your data into a central place and document the project. To complete the exercise, gather all of the data from a project and work through the steps to prepare the data for future reuse. This exercise refers to several other exercises in the Workbook that should be completed during this process, if they have not been already.

This exercise was adapted from the "Project Close-Out Checklist" (Briney 2020c).

Prepare data:
- ☐ Move all data into one central project folder; this folder may have sub-folders and should be organized however makes sense for your data (see Exercise 3.1: Set up a file organization system).
- ☐ As necessary, work through Exercise 7.1: Convert data file types to copy data into more open/common file formats.
- ☐ Perform any necessary quality control on the data.

Back up your research notes:
- ☐ If your notes are electronic, save a copy in the project folder.
- ☐ If your notes are physical, scan them and save a copy in the project folder.

Create a project archive folder:
- ☐ Work through Exercise 7.2: Create an "archive" folder. Put the archive folder in the project folder.

Document the data:
- ☐ If you haven't done so already, work through Exercise 2.2: Document project information. Store a copy of this file in the top level of your central project folder.
- ☐ Record citations to any publications resulting from the project and permanent identifiers, such as DOIs, for any publicly available datasets.

Save files in a stable location:
- ☐ Save the project folder on a storage system that will be accessible for the next several years.
- ☐ Back up the files in a reliable backup system.

Exercise 7.4: Separate from Your Institution

Researchers regularly leave institutions in order to take up new jobs. For how common this occurrence is, it represents a critical, often busy transition period during which data may be lost. It's therefore helpful to have a list of all of the tasks to do lest something is forgotten. While this exercise doesn't cover everything you might need to do when leaving your current institution, it does focus on the tasks necessary for making sure you maintain access to the data that you are entitled to keep.

This checklist outlines a number of important steps that researchers should take to ensure that they retain the appropriate data yet leave behind what belongs to the institution. The researcher leaving the institution should work through this checklist, potentially with the project lead, to ensure they keep the proper information while returning what does not belong to them.

This exercise was adapted from the "Data Departure Checklist" (Goben and Briney 2023).

Retain copies of data that you have permission to keep:
- ☐ If you have not done so already, work through Exercise 5.4: Determine data stewardship to determine what data you may retain.
- ☐ Identify and keep pertinent research data from personal computers and devices (e.g. external hard drives).
- ☐ Identify and keep pertinent research data from storage systems (e.g. AWS/Azure, Box, campus HPC, Dropbox, Electronic Lab Notebook, Globus, Google Drive, group/department/college servers, Microsoft OneDrive, Microsoft Sharepoint, or shared collaborator drives).
- ☐ When appropriate, make a copy of research notes.

Delete personal information and remove personal devices:
- ☐ Remove personal information from research group computers and devices.
- ☐ Remove personal computers and devices from research group spaces.
- ☐ Remove personal access to shared accounts (e.g. research group GitHub, research group website, mailing lists, or social media).

Return research group hardware:
- ☐ Individual computer/workstation.
- ☐ Tablet(s).
- ☐ Peripherals (e.g. keyboard, mouse, monitor).
- ☐ External drives.
- ☐ Other research group equipment (e.g. cameras, recording devices).

Update research administration documents, as necessary:

☐ Update/transfer Institutional/Ethical Review Board.

☐ Update/transfer animal care paperwork.

☐ Update/transfer Data Use Agreements (DUA).

☐ Update/transfer Material Transfer Agreements (MTA).

☐ Update/transfer research grants.

Handle email:

☐ Set out of office, providing forwarding information.

☐ Forward/backup important emails.

☐ Check with the institute Archivist or Records Manager for retention policies (depends on rank).

REFERENCES

APA (2020) *Data set references*. Available at: https://apastyle.apa.org/style-grammar-guidelines/references/examples/data-set-references (Accessed: 16 January 2025).

Barbosa, S. et al. (2024) 'Generalist Repository Selection Flowchart'. Zenodo. Available at: https://doi.org/10.5281/zenodo.11105430

Briney, K. et al. (2022) 'Data Doubles Data Management Plans'. OSF. Available at: https://doi.org/10.17605/OSF.IO/JE7QP (Accessed: 12 May 2025).

Briney, K.A. (2015) *Data Management for Researchers: Organize, Maintain and Share Your Data for Research Success*. Exeter, UK: Pelagic Publishing.

Briney, K.A. (2020a) 'Data Management Plan (DMP) Checklist'. California Institute of Technology. Available at: https://doi.org/10.7907/rksd-4x31 (Accessed: 12 May 2025).

Briney, K.A. (2020b) 'File Naming Convention Worksheet'. California Institute of Technology. Available at: https://doi.org/10.7907/894q-zr22 (Accessed: 12 May 2025).

Briney, K.A. (2020c) 'Project Close-Out Checklist for Research Data'. California Institute of Technology. Available at: https://doi.org/10.7907/yjph-sa32 (Accessed: 12 May 2025).

Briney, K.A. (2023) 'Leveling Up Data Management'. California Institute of Technology. Available at: https://doi.org/10.7907/syk7-3z92 (Accessed: 12 May 2025).

Briney, K.A. (2024) 'Measuring data rot: An analysis of the continued availability of shared data from a Single University', *PLoS ONE*, 19(6), p. e0304781. Available at: https://doi.org/10.1371/journal.pone.0304781

Broman, K.W. and Woo, K.H. (2018) 'Data Organization in Spreadsheets', *The American Statistician*, 72(1), pp. 2–10. Available at: https://doi.org/10.1080/00031305.2017.1375989

Cesal, A. (2021) 'Writing Alt Text for Data Visualization', *Nightingale*, 8 June. Available at: https://nightingaledvs.com/writing-alt-text-for-data-visualization/ (Accessed: 9 July 2024).

Creative Commons Wiki (2014) *CC0 use for data*. Available at: https://wiki.creativecommons.org/wiki/CC0_use_for_data (Accessed: 18 August 2023).

Curtin, L. et al. (2025) 'Best Practices for Data Submission in Generalist Repositories: A Checklist'. Zenodo. Available at: https://zenodo.org/records/14278907 (Accessed: 18 March 2025).

DIAGRAM Center (2019) *Image Description Guidelines, DIAGRAM Center*. Available at: http://diagramcenter.org/table-of-contents-2.html/ (Accessed: 19 July 2024).

Electronic Frontier Foundation (2016a) *EFF Dice-Generated Passphrases*. Available at: https://www.eff.org/dice (Accessed: 16 January 2025).

Electronic Frontier Foundation (2016b) *EFF's Long Wordlist*. Available at: https://www.eff.org/files/2016/07/18/eff_large_wordlist.txt (Accessed: 16 January 2025).

EPSRC (2024) *Policy framework on research data*. Available at: https://www.ukri.org/who-we-are/epsrc/our-policies-and-standards/policy-framework-on-research-data/ (Accessed: 24 January 2025).

European Commission (2024) *Open Science*. Available at: https://research-and-innovation.ec.europa.eu/strategy/strategy-research-and-innovation/our-digital-future/open-science_en (Accessed: 24 January 2025).

FASEB (2021) *DataWorks! Data Management Plan Challenge*. Available at: https://www.faseb.org/data-management-and-sharing/dataworks-prize/dataworks-dmp-challenge (Accessed: 17 March 2025).

GBH (2024) *Guidelines for Describing STEM Images, GBH*. Available at: https://www.wgbh.org/guidelines-for-describing-stem-images (Accessed: 19 July 2024).

Goben, A. and Briney, K.A. (2023) 'Data Departure Checklist'. Available at: https://doi.org/10.7907/h314-4x51 (Accessed: 11 August 2023).

Government of Canada (2021) *Tri-Agency Research Data Management Policy*. Innovation, Science and Economic Development Canada. Available at: https://science.gc.ca/site/science/en/interagency-research-funding/policies-and-guidelines/research-data-management/tri-agency-research-data-management-policy (Accessed: 16 January 2025).

Hearn, M. (2016) *Use a Passphrase*. Available at: https://www.useapassphrase.com/ (Accessed: 15 January 2025).

IASSIST SIGDC (2012) *Quick Guide to Data Citation*. Available at: https://www.icpsr.umich.edu/files/ICPSR/enewsletters/iassist.html (Accessed: 4 March 2025).

Kincaid, E. (2023) 'Spider researcher Jonathan Pruitt faked data in multiple papers, university finds', *Retraction Watch*, 10 May. Available at: https://retractionwatch.com/2023/05/10/spider-researcher-jonathan-pruitt-faked-data-in-multiple-papers-university-finds/ (Accessed: 9 January 2025).

Kovacs, M., Hoekstra, R. and Aczel, B. (2021) 'The Role of Human Fallibility in Psychological Research: A Survey of Mistakes in Data Management', *Advances in Methods and Practices in Psychological Science*, 4(4), p. 25152459211045930. Available at: https://doi.org/10.1177/25152459211045930

Laskowski, K. (2020) 'What to do when you don't trust your data anymore', *Laskowski Lab at UC Davis*, 29 January. Available at: https://laskowskilab.faculty.ucdavis.edu/2020/01/29/retractions/ (Accessed: 9 January 2025).

Laskowski, K. (2024) 'What to Do When You Don't Trust Your Data Anymore'. *2024 Conference of the Society for Open, Reliable, and Transparent Ecology and Evolutionary Biology*, 15 October. Available at: https://www.youtube.com/watch?v=xxDeqZyxpgc (Accessed: 9 January 2025).

Laskowski, K.L., Montiglio, P.-O. and Pruitt, J.N. (2016) 'Retracted: Individual and Group Performance Suffers from Social Niche Disruption', *The American Naturalist*, 187(6), pp. 776–785. Available at: https://doi.org/10.1086/686220

Laskowski, K.L., Montiglio, P.-O. and Pruitt, J.N. (2020) 'Retraction', *The American Naturalist*, 195(2), pp. 393–393. Available at: https://doi.org/10.1086/708066

López Lloreda, C. (2023) 'University investigation found prominent spider biologist fabricated, falsified data', *Science*, 11 May. Available at: https://doi.org/10.1126/science.adi6906 (Accessed: 9 January 2025).

Marcus, A. (2020a) 'Authors questioning papers at nearly two dozen journals in wake of spider paper retraction', *Retraction Watch*, 29 January. Available at: https://retractionwatch.com/2020/01/29/authors-questioning-papers-at-nearly-two-dozen-journals-in-wake-of-spider-paper-retraction/ (Accessed: 9 January 2025).

Marcus, A. (2020b) '"I'm starting the year off with something I didn't expect to ever do: I'm retracting a paper."', *Retraction Watch*, 20 January. Available at: https://retractionwatch.com/2020/01/20/im-starting-the-year-off-with-something-i-didnt-expect-to-ever-do-im-retracting-a-paper/ (Accessed: 9 January 2025).

McCarty, N. (2021a) 'Bad blood at a lab leads to retraction after postdoc publishes study without supervisor's permission', *Retraction Watch*, 15 March. Available at: https://

retractionwatch.com/2021/03/15/bad-blood-at-a-lab-leads-to-retraction-after-postdoc-publishes-study-without-supervisors-permission/ (Accessed: 17 March 2025).

McCarty, N. (2021b) 'Journal pulls two studies that listed an author without his permission', *Retraction Watch*, 1 February. Available at: https://retractionwatch.com/2021/02/01/journal-pulls-two-studies-that-listed-an-author-without-his-permission/ (Accessed: 17 March 2025).

McMaster University (2023) *Update on research misconduct case, Daily News*. Available at: https://dailynews.mcmaster.ca/worthmentioning/update-on-research-misconduct-case/ (Accessed: 9 January 2025).

Murray-Rust, Peter et al. (2010) *Panton Principles*. Available at: https://pantonprinciples.okfn.org/index.html (Accessed: 18 February 2025).

NASA et al. (2024) *Galaxies IC 2163 and NGC 2207 (Webb MIRI Image), Webb*. Available at: https://webbtelescope.org/contents/media/images/2024/136/01J9S544K01B5TSTPX5MS5PCNA (Accessed: 3 February 2025).

NIH Office of Data Science Strategy (2024) *Announcing the 2023 DataWorks! Prize Winners*. Available at: https://datascience.nih.gov/director/directors-blog-dataworks-winners-2024 (Accessed: 14 January 2025).

NIST (2025) *How Do I Create a Good Password?* Available at: https://www.nist.gov/cybersecurity/how-do-i-create-good-password (Accessed: 2 May 2025).

Open Knowledge Foundation (2025) *Open Data Commons: legal tools for open data*. Available at: https://opendatacommons.org/ (Accessed: 18 February 2025).

ORCID (2025) *ORCID*. Available at: https://orcid.org/ (Accessed: 9 May 2025).

Oxford, E. and Woodbrook, R. (2023) *Accessibility Data Curation Primer*. Available at: https://github.com/DataCurationNetwork/data-primers/blob/main/Accessibility%20Data%20Curation%20Primer/accessibility-data-curation-primer.md (Accessed: 30 April 2024).

Patrias, K. and Wendling, D. (2018) 'Databases/Retrieval Systems/Datasets on the Internet', in *Citing Medicine: The NLM Style Guide for Authors, Editors, and Publishers [Internet]. 2nd edition*. National Library of Medicine (US). Available at: https://www.ncbi.nlm.nih.gov/books/NBK7273/ (Accessed: 9 May 2025).

PLoS ONE (2023) *Recommended Repositories*. Available at: https://journals.plos.org/plosone/s/recommended-repositories (Accessed: 12 May 2025).

Purdue University (2024) *Purdue Online Writing Lab*. Available at: https://owl.purdue.edu/owl/index.html (Accessed: 5 March 2025).

re3data (2025) *Registry of Research Data Repositories*. Available at: https://www.re3data.org/ (Accessed: 5 February 2025).

Schwabish, J., Popkin, S.J. and Feng, A. (2022) *Do No Harm Guide: Centering Accessibility in Data Visualization*. Available at: https://www.urban.org/research/publication/do-no-harm-guide-centering-accessibility-data-visualization (Accessed: 30 September 2024).

Springer Nature (2025) *Recommended repositories*. Available at: https://www.springernature.com/gp/authors/research-data-policy/recommended-repositories (Accessed: 11 March 2025).

Tenopir, C. et al. (2016) 'Data Management Education from the Perspective of Science Educators', *International Journal of Digital Curation*, 11(1), pp. 232–251. Available at: https://doi.org/10.2218/ijdc.v11i1.389

The Center for Scientific Integrity (2025) 'Retraction Watch Database'. Available at: http://retractiondatabase.org/RetractionSearch.aspx? (Accessed: 9 January 2025).

Vines, T.H. et al. (2014) 'Data from: The availability of research data declines rapidly with article age'. Dryad. Available at: https://doi.org/10.5061/DRYAD.Q3G37

Vogel, T. et al. (2025) 'ACS Style Quick Guide', in *The ACS Guide to Scholarly Communication*. American Chemical Society (ACS Guide to Scholarly Communication). Available at: https://doi.org/10.1021/acsguide.40303

Vrouwenvelder, K., Raia, N.H. and Thomer, A.K. (2025) 'Obstacles to Dataset Citation Using Bibliographic Management Software', *Data Science Journal*, 24. https://doi.org/10.5334/dsj-2025-017

White House Office of Science and Technology Policy (OSTP) (2022) *Ensuring Free, Immediate, and Equitable Access to Federally Funded Research*. Executive Office of the President of the United States. Available at: https://bidenwhitehouse.archives.gov/wp-content/uploads/2022/08/08-2022-OSTP-Public-Access-Memo.pdf (Accessed: 12 May 2025).

Wickham, H. (2014) 'Tidy Data', *Journal of Statistical Software*, 59, pp. 1–23. Available at: https://doi.org/10.18637/jss.v059.i10

Wilkinson, M.D. et al. (2016) 'The FAIR Guiding Principles for scientific data management and stewardship', *Scientific Data*, 3(1), p. 160018. Available at: https://doi.org/10.1038/sdata.2016.18

Zehr, S.M. et al. (2019) 'Data from: Life history profiles for 27 strepsirrhine primate taxa generated using captive data from the Duke Lemur Center'. Dryad. Available at: https://doi.org/10.5061/DRYAD.FJ974